FERTIGATION
Fundamentals & Applications

Contents

1

Introduction

Fertigation is the application of fertilizers, soil amendments, or other water soluble products through an irrigation system. Chemigation, a related and sometimes interchangeable term, is the application of chemicals through an irrigation system. Chemigation is considered to be a more restrictive and controlled process due to the potential nature of the products being delivered (pesticides, herbicides, fungicides, etc.) to cause harm to humans, animals or the environment. Therefore, chemigation is generally more regulated than fertigation.

Fertigation in Use

Fertigation is used extensively in commercial agriculture and horticulture and is starting to be used in general landscape applications as dispenser units become more reliable and easy to use:

- Fertigation is used to spoon feed additional nutrients or correct nutrient deficiencies detected in plant tissue analysis.
- Usually practiced with high value crops such as vegetables, turf, fruit trees, and ornamentals.
- Injection during middle one-third or the middle one-half of the irrigation recommended for fertigation using micro propagation.

- Keep water supply for fertigation separate from domestic water supply to avoid contamination.
- Allow you to change your fertilizer programme during the growing season in order to adjust for fruit, flower, and root development.

Methods Used in Fertigation

- Drip irrigation which reduces per water and nutrient application rates relative to sprinklers.
- Sprinkler systems increase leaf and fruit quality.
- Other methods of application include lateral move, traveller gun, and solid set systems.
- All systems should be placed on a raised and/or sealed platform, not in direct contact with earth.
- Fitted with chemical spill trays.
- Continuous application- fertilizer is supplied at a constant rate.
- Three stage application-irrigation starts without fertilizers and then the later in process fertilizers are applied.
- Proportional application-injection rate is proportional to water discharge rate.
- Quantitative application-nutrient solution is applied in a calculated amount to each irrigation block.
- To determine the injection rate for the particular fertilizer being used use the formula: Maximum injection rate = $(5 \times Q \times L) / (f \times 60)$.

where,

Q = irrigation pump discharge in litres per second.

L = fertilizer tank volume in litres.

F = amount of fertilizer in grams.

System Design

Fertigation assists distribution of fertilizers for farmers. The simplest type of fertigation system consists of a tank with a pump, distribution pipes, capillaries and dripper pen.

What should be Considered ?

- Water quality;
- Soil type;
- Nutrient consumption (daily);
- Appropriate nutrient materials.

Possible Strategies to be used:

- Injecting for short time periods at the beginning, middle, and end of irrigation cycle.
- Injecting during middle 50% of the irrigation cycle.
- Continuous irrigation.

Commonly Used Nutrients

- Most plant nutrients can be applied through irrigation systems.
- Nitrogen is most commonly used nutrient.
- Other nutrients include nitrate, ammonium, urea, phosphate and potassium.

Determining which Nutrient is Used

- A soil fertility analysis is used to determine which of the more stable nutrients should be used.

What's Good about it?

Benefits of fertigation over traditional broadcast or drop fertilizing methods include:

- Increased nutrient absorption by plants.
- Reduction in fertilizer and chemicals needed.
- Reduced leaching to the water table and,
- Reduction in water usage due to the plant's resulting increased root mass being able to trap and hold water.
- Application of nutrients at the precise time they are needed and at the rate they are utilised.

Disadvantages

- Concentration of solution decreases as fertilizer dissolves, leading to poor nutrient placement.

- Requires pressure loss in main irrigation line.
- Limited capacity.

Controls

Because of the potential risk in contaminating the potable (drinking) water supply, a backflow prevention device is required for most fertigation systems. Backflow requirements vary greatly so it is very important to understand the proper level of backflow prevention required by law. In the United States the minimum backflow protection is usually determined by state regulation but each city or town may increase the level of protection required.

Deficit Irrigation

Deficit irrigation (DI) has been reviewed and defined as follows:

> "Deficit irrigation is an optimisation strategy in which irrigation is applied during drought-sensitive growth stages of a crop. Outside these periods, irrigation is limited or even unnecessary if rainfall provides a minimum supply of water. Water restriction is limited to drought-tolerant phenological stages, often the vegetative stages and the late ripening period. Total irrigation application is therefore not proportional to irrigation requirements throughout the crop cycle. While this inevitably results in plant drought stress and consequently in production loss, DI maximises water productivity, which is the main limiting factor (English, 1990). In other words, DI aims at stabilising yields and at obtaining maximum crop water productivity rather than maximum yields (Zhang and Oweis,1999)."

The Concept

DI is sometimes referred to as incomplete supplemental irrigation or regulated DI. It is a watering strategy that can be applied by different types of irrigation application methods.

The correct application of DI requires thorough understanding of the yield response to water and of the economic impact of reductions in harvest. In regions where water resources are restrictive it can be more profitable for a farmer to maximise crop water productivity in stead of maximising the harvest per unit land. The saved water can be used for other purposes or to irrigate extra units of land.

Crop Water Productivity

Crop 'Water Productivity' (WP) or 'Water Use Efficiency' (WUE) expressed in kg/cu m is an efficiency term, expressing the amount of marketable product (e.g. kilograms of grain) in relation to the amount of input needed to produce that output (cubic metres of water). The water used for crop production is referred to as crop evapotranspiration. This is a combination of water lost by evaporation from the soil surface and transpiration by the plant, occurring simultaneously. Except by modelling, distinguishing between the two processes is difficult. Representative values of WUE for cereals at field level, expressed with evapotranspiration in the denominator, can vary between 0.10 and 4 kg/cu m.

Experiences with Deficit Irrigation

For certain crops, experiments confirm that DI can increase water use efficiency without severe yield reductions. For example for winter wheat in Turkey, planned DI increased yields by 65% as compared to winter wheat under rain-fed cultivation, and had double the water use efficiency as compared to rain-fed and fully irrigated winter wheat. Similar positive results have been described for cotton. Experiments in Turkey and India indicated that the irrigation water use for cotton could be reduced to up to 60 percent of the total crop water requirement without severe yield losses. In this way, high water productivity and a better nutrient-water balance was obtained.

Certain Underutilised and horticultural crops also respond favourably to DI, such as tested at experimental and farmer level for the crop quinoa. Yields could be stabilised at around

1.6 tons per hectare by supplementing irrigation water if rainwater was lacking during the plant establishment and reproductive stages. Applying irrigation water throughout the whole season (full irrigation) reduced the water productivity. Also in viticulture and fruit tree cultivation, DI is a common practice.

For other crops, the application of deficit irrigation will result in a lower water use efficiency and yield. This is the case when crops are sensitive to drought stress throughout the season, such as maize.

Reasons for Increased Water Productivity under Deficit Irrigation

From worldwide literature on DI, it is clear that DI is able to increase the ratio of yield over crop water consumption (evapotranspiration) by reducing the water loss by unproductive evaporation, and/or by increasing the proportion of marketable yield to the totally produced biomass (harvest index), and/or by increasing the proportion of total biomass production to transpiration due to hardening of the crop - although this effect is very limited due to the conservative relation between biomass production and crop transpiration - and/or due to adequate fertilizer application, and/or by avoiding bad agronomic conditions during crop growth, such as water logging in the root zone, pests and diseases, etc.

Advantages

The correct application of DI for a certain crop:

- maximises the productivity of water with good harvest quality;
- allows economic planning and stable income due to a stabilisation of the harvest in comparison with rain-fed cultivation;
- decreases the risk of certain diseases linked to high humidity (e.g. fungi) in comparison with full irrigation;
- reduces nutrient loss by leaching of the root zone, which results in better groundwater quality and lower fertilizer needs as for cultivation under full irrigation;

- improves control over the sowing date and length of the growing period independent from the onset of the rainy season and therefore improves agricultural planning.

Constraints

Apart from all the advantages, a number of constraints can be listed:

- Exact knowledge of the crop response to water stress is imperative.
- There should be sufficient flexibility in access to water during periods of high demand (drought sensitive stages of a crop).
- A minimum quantity of water should be guaranteed for the crop, below which DI has no significant beneficial effect.
- An individual farmer should consider the benefit for the total water users community (extra land can be irrigated with the saved water), when he faces a below-maximum yield.
- Because irrigation is applied more efficiently, the risk for soil salinisation is somewhat higher under DI as compared to full irrigation.

Growth in Fertigation

Fertigation has increased dramatically in the past 15 years, particularly for sprinkler and drip systems. For drip systems, the expansion is mostly in horticultural and high value crops. In agricultural areas with declining water supplies, drip systems have also increased. With increased irrigation, a corresponding increase in fertigation has taken place. It will continue to grow since such systems result in less water usage and better uniformity and lend themselves to the technique much more readily than the less water-efficient and non-uniform furrow and flood systems being replaced.

Effective fertigation requires knowledge of certain plant characteristics such as optimum daily nutrient consumption rate and root distribution in the soil. Nutrient characteristics such as solubility and mobility are important and irrigation water quality factors such as pH, mineral content, salinity and nutrient solubility must be considered.

The need for irrigation is the main factor in fertigation because the irrigation system is primarily installed to provide water. The opportunity to fertigate is an added benefit. This section will address issues that must be considered to achieve maximum benefits when fertigating.

Important Factors to Consider when Fertigating

- Distribution uniformity of the irrigation water is critical for uniform application of the fertilizer.
- Surface irrigation systems generally produce surface run-off. Make sure that any surface run-off during fertigation is picked up and reused on the field being treated or a similar field that is also being treated. Fertigation may not be suitable on fields that are very susceptible to run-off (i.e. steep slopes and low infiltration rates).
- Make sure the combination of fertilizers and water will not produce clogging if using a microirrigation system.
- Know how much water is needed to refill the root zone during the irrigation and plan the fertilizer application accordingly. Over-irrigation during fertigation not only wastes water but could result in leaching of the fertilizer below the root zone.
- The type of injection device is critical depending on the type of irrigation system being used. Some devices will inject at a relatively uniform rate throughout the irrigation, some will not. Be aware of which type is being used and which type is required by the situation.
- Be aware of the requirements for flushing the irrigation system after fertigation. This may take 10 to 15 minutes. Also run clean water through the injection metre, discharge hose and check valve.

- Irrigation systems should be monitored much more closely during fertigation, continuously if possible.
- Under saline conditions, salinity problems can be intensified by fertigation and improper irrigation management. The level of salinity in soil may itself be increased, even temporarily, by fertilizer application as mineral fertilizers are for the most part soluble salts.

Summary

When the advantages of fertigation are exploited and the disadvantages are minimised, fertigation can be a very efficient method of fertilization. Although not an end in itself, fertigation can compliment a good base soil fertility programme. An understanding of the types of soil, water quality, appropriate nutrient materials, safety precautions, and cropping systems are all important factors in making fertigation useful and economical.

2

Nutrient Management

A nutrient is a chemical that an organism needs to live and grow or a substance used in an organism's metabolism which must be taken in from its environment. Methods for nutrient intake vary, with animals and protists consuming foods that are digested by an internal digestive system, but most plants ingest nutrients directly from the soil through their roots or from the atmosphere. Some plants, like carnivorous plants, externally digest nutrients from animals, before ingesting them. The effects of nutrients are dose-dependent.

Organic nutrients include carbohydrates, fats, proteins (or their building blocks, amino acids), and vitamins. Inorganic chemical compounds such as dietary minerals, water, and oxygen may also be considered nutrients. A nutrient is essential to an organism if it cannot be synthesised by the organism in sufficient quantities and must be obtained from an external source. Nutrients needed in relatively large quantities are called macronutrients and those needed in relatively small quantities are called micronutrients.

Types of Nutrients

Macronutrients are defined in several different ways.

- The chemical elements humans consume in the largest quantities are carbon, hydrogen, nitrogen, oxygen, phosphorus, and sulphur.

- The classes of chemical compounds humans consume in the largest quantities and which provide bulk energy are carbohydrates, proteins, and fats. Water and atmospheric oxygen also must be consumed in large quantities, but are not always considered "food" or "nutrients".
- Calcium, salt (sodium and chloride), magnesium, and potassium (along with phosphorus and sulphur) are sometimes added to the list of macronutrients because they are required in large quantities compared to other vitamins and minerals. They are sometimes referred to as the macrominerals.

The remaining vitamins, minerals, fats or elements, are called micronutrients because they are required in relatively small quantities.

Substances, Providing Energy

- Carbohydrates are compounds made up of sugars. Carbohydrates are classified by their number of sugar units: monosaccharides (such as glucose and fructose), disaccharides (such as sucrose and lactose), oligosaccharides, and polysaccharides (such as starch, glycogen, and cellulose).
- Proteins are organic compounds that consists of the amino acids joined by peptide bonds. The body cannot manufacture some of the amino acids (termed essential amino acids); the diet must supply these. In nutrition, proteins are broken down through digestion by proteases back into free amino acids.
- Fats consist of a glycerin molecule with three fatty acids attached. Fatty acids are unbranched hydrocarbon chains, connected by single bonds alone (saturated fatty acids) or by both double and single bonds (unsaturated fatty acids). Fats are needed to keep cell membranes functioning properly, to insulate body organs against shock, to keep body temperature stable, and to maintain healthy skin and hair. The body does not manufacture

certain fatty acids (termed essential fatty acids) and the diet must supply these.

- Fat has an energy content of 9 kcal/g (~37.7 kJ/g); proteins and carbohydrates 4 kcal/g (~16.7 kJ/g). Ethanol (grain alcohol) has an energy content of 7 kcal/g (~29.3 kJ/g).

Substances, Supporting Metabolism

- Dietary minerals are generally trace elements, salts, or ions such as copper and iron. Some of these minerals are essential to human metabolism.
- Vitamins are organic compounds essential to the body. They usually act as coenzymes or cofactors for various proteins in the body.
- Water is an essential nutrient and is the solvent in which all the chemical reactions of life take place.

Nutrients and Plants

The chemical elements consumed in the greatest quantities by plants are carbon, hydrogen, and oxygen. These are present in the environment in the form of water and carbon dioxide; energy is provided by sunlight. Nitrogen, phosphorus, potassium, and sulphur are also needed in relatively large quantities. Together, these are the elemental macronutrients for plants, often represented by the acronym CHNOPS. Usually they are sourced from inorganic (e.g. carbon dioxide, water, nitrate, phosphate, sulphate) or organic (e.g. carbohydrates, lipids, proteins) compounds, although elemental diatomic molecules of nitrogen and (especially) oxygen are often used.

Other chemical elements are also necessary to carry out various life processes and build structures.

Some of these are considered macronutrients in certain organisms. The acronym C. HOPKiN'S CaFe Mg (to be used as C. Hopkins coffee mug) is used by some students to remember the list as: Carbon, Hydrogen, Oxygen, Phosphorus, Potassium (K), Nitrogen, Sulphur, Calcium, Iron (Fe), and Magnesium (Mg). Silicon, chloride, sodium, copper, zinc, and molybdenum are sometimes also included, but are in other cases considered micronutrients.

Oversupply of plant nutrients in the environment can cause excessive plant and algae growth. Eutrophication, as this process is called, may cause imbalances in population numbers and other nutrients that can be harmful to certain species. For example, an algal bloom can deplete the oxygen available for fish to breathe. Causes include water pollution from sewage or run-off from farms (carrying excess agricultural fertilizer). Nitrogen and phosphorus are most commonly the limiting factor in growth, and thus the most likely to trigger eutrophication when introduced artificially.

Essential and Non-essential Nutrients

Nutrients are frequently categorised as essential or nonessential. Essential nutrients are unable to be synthesised internally (either at all, or in sufficient quantities), and so must be consumed by an organism from its environment.

For humans, these include essential fatty acids, essential amino acids, vitamins, and certain dietary minerals. Oxygen and water are also essential for human survival, but are generally not considered "food" when consumed in isolation.

Humans can derive energy from a wide variety of fats, carbohydrates, proteins, and ethanol, and can synthesise other needed amino acids from the essential nutrients.

Non-essential nutrients can still have a significant impact on health, whether beneficial or toxic. For example, most dietary fibre is not absorbed by the human digestive tract, but is important in digestion and absorption of otherwise harmful substances. Interest has recently increased in phytochemicals, which include many non-essential nutrients which may have health benefits.

Essential Nutrients

An essential nutrient is a nutrient required for normal body functioning that either cannot be synthesised by the body at all, or cannot be synthesised in amounts adequate for good health (e.g. niacin, choline), and thus must be obtained from a dietary source. Some categories of essential nutrients include vitamins, dietary minerals, essential fatty acids, and essential amino acids.

Different species have very different essential nutrients. For example, most mammals synthesise their own ascorbic acid, and it is therefore not considered an essential nutrient for such species. It is, however, an essential nutrient for human beings, who require external sources of ascorbic acid (known as Vitamin C in the context of nutrition).

Many essential nutrients are toxic in large doses. Some can be taken in amounts larger than required in a typical diet, with no apparent ill effects. Linus Pauling said of vitamin B_3, (either niacin or niacinamide), "What astonished me was the very low toxicity of a substance that has such very great physiological power. A little pinch, 5 mg, every day, is enough to keep a person from dying of pellagra, but it is so lacking in toxicity that ten thousand times as much can [sometimes] be taken without harm."

List of Essential Nutrients

- *Essential human requirements usually not considered to be nutrients:*
  - Oxygen;
  - Water;
  - Sunlight (for synthesis of Vitamin D).
- *Essential fatty acids:*
  - Linolenic acid (the shortest chain omega-3 fatty acid);
  - Linoleic acid (the shortest chain omega-6 fatty acid).
- *Essential amino acids necessary for all humans:*
  - Isoleucine;
  - Lysine;
  - Leucine;
  - Methionine;

- Phenylalanine;
- Threonine;
- Tryptophan;
- Valine.

- *Essential amino acids necessary for human children but not adults:*
  - Histidine;
  - Arginine.
- *Vitamins:*
  - Biotin (vitamin B7, vitamin H);
  - Choline (vitamin Bp);
  - Folate (folic acid, vitamin B9, vitamin M);
  - Niacin (vitamin B3, vitamin P, vitamin PP);
  - Pantothenic acid (vitamin B5);
  - Riboflavin (vitamin B2, vitamin G);
  - Thiamine (vitamin B1);
  - Vitamin A (retinol);
  - Vitamin B6 (pyridoxine, pyridoxamine, or pyridoxal);
  - Vitamin B12 (cobalamin);
  - Vitamin C (ascorbic acid);
  - Vitamin D (ergocalciferol, or cholecalciferol);
  - Vitamin E (tocopherol);
  - Vitamin K (naphthoquinoids).
- *Dietary minerals:*
  - Calcium (Ca);
  - Chloride (Cl^-);
  - Cobalt (Co) (as part of Vitamin B-12);
  - Copper (Cu);
  - Iodine (I);

- Iron (Fe);
- Magnesium (Mg);
- Manganese (Mn);
- Molybdenum (Mo);
- Nickel (Ni);
- Phosphorus (P);
- Potassium (K);
- Selenium (Se);
- Sodium (Na);
- Sulphur (S) numerous roles;
- Zinc (Zn).

The body's requirements vary widely. At one extreme a 70 kg human contains 1.0 kg of calcium but only 3 mg of cobalt.

Elements with Speculated Role in Human Health

Many elements have been implicated at various times to have a role in human health. For none of these elements has a specific protein or complex been identified:

- Boron (B);
- Chromium (Cr);
- Silicon (Si);
- Arsenic (As).

Plant Nutrition

Plant nutrition is the study of the chemical elements that are necessary for plant growth. There are several principles that apply to plant nutrition. Some elements are directly involved in plant metabolism (Arnon and Stout, 1939). However, this principle does not account for the so-called beneficial elements, whose presence, while not required, has clear positive effects on plant growth.

A nutrient that is able to limit plant growth according to Liebig's law of the minimum, is considered an essential plant nutrient if the plant can not complete its full life cycle without it. There are 14 essential plant nutrients.

Carbon, Hydrogen and Oxygen are essential for plants, but are not considered plant nutrients.

Macronutrients:

- N = Nitrogen
- K = Potassium
- Ca = Calcium
- P = Phosphorus
- Mg = Magnesium
- S = Sulphur
- Si = Silicon

Micronutrients (trace levels) include:

- Cl = Chlorine
- Fe = Iron
- B = Boron
- Mn = Manganese
- Na = Sodium
- Zn = Zinc
- Cu = Copper
- Ni= Nickel
- Mo = Molybdenum

These nutrients are further divided into the mobile and immobile nutrients. A plant will always supply more nutrients to its younger leaves than its older ones, so when nutrients are mobile, the lack of nutrients is first visible on older leaves. When a nutrient is less mobile, the younger leaves suffer because the nutrient does not move up to them but stays lower in the older leaves. Nitrogen, phosphorus, and potassium are mobile nutrients, while the others have varying degrees of mobility.

Processes

Plants uptake essential elements from the soil through their roots and from the air through their leaves. Nutrient uptake in the soil is achieved by cation exchange, wherein root hairs pump hydrogen ions (H^+) into the soil through proton pumps. These hydrogen ions displace cations attached to negatively charged soil particles so that the cations are available for uptake by the root. In the leaves, stomata open to take in carbon dioxide and expel oxygen. The carbon dioxide molecules are used as the carbon source in photosynthesis.

Though nitrogen is plentiful in the earth's atmosphere, relatively few plants engage in nitrogen fixation (conversion of atmospheric nitrogen to a biologically useful form). Most plants therefore require nitrogen compounds to be present in the soil in which they grow.

Functions of Nutrients

Each of these nutrients is used in a different place for a different essential function.

Carbon: Carbon forms the backbone of many plants biomolecules, including starches and cellulose. Carbon is fixed through photosynthesis from the carbon dioxide in the air and is a part of the carbohydrates that store energy in the plant.

Hydrogen: Hydrogen also is necessary for building sugars and building the plant. It is obtained almost entirely from water.

Oxygen: Oxygen is necessary for cellular respiration. Cellular respiration is the process of generating energy-rich adenosine triphosphate (ATP) via the consumption of sugars made in photosynthesis. Oxygen gas is produced as a by-product from this reaction.

Phosphorus: Phosphorus is important in plant bioenergetics. As a component of ATP, phosphorus is needed for the conversion of light energy to chemical energy (ATP) during photosynthesis. Phosphorus can also be used to modify the activity of various enzymes by phosphorylation, and can be used for cell signalling. Since ATP can be used for the biosynthesis of many plant biomolecules, phosphorus is important for plant growth and flower/seed formation.

Potassium: Potassium regulates the opening and closing of the stoma by a potassium ion pump. Since stomata are important in water regulation, potassium reduces water loss from the leaves and increases drought tolerance. Potassium deficiency may cause necrosis or interveinal chlorosis.

Nitrogen: Nitrogen is an essential component of all proteins. Nitrogen deficiency most often results in stunted growth.

Sulphur: Sulphur is a structural component of some amino acids and vitamins, and is essential in the manufacturing of chloroplasts.

Silicon: Silicon is deposited in cell walls and contributes to its mechanical properties including rigidity and elasticity

Calcium: Calcium regulates transport of other nutrients into the plant and is also involved in the activation of certain plant enzymes. Calcium deficiency results in stunting.

Magnesium: Magnesium is an important part of chlorophyll, a critical plant pigment important in photosynthesis. It is important in the production of ATP through its role as an enzyme cofactor. There are many other biological roles for magnesium. Magnesium deficiency can result in interveinal chlorosis.

Iron: Iron is necessary for photosynthesis and is present as an enzyme cofactor in plants. Iron deficiency can result in interveinal chlorosis and necrosis.

Molybdenum: Molybdenum is a cofactor to enzymes important in building amino acids.

Boron: Boron is important in sugar transport, cell division, and synthesising certain enzymes. Boron deficiency causes necrosis in young leaves and stunting.

Copper: Copper is important for photosynthesis. Symptoms for copper deficiency include chlorosis. Involved in many enzyme processes. Necessary for proper photosythesis. Involved in the manufacture of lignin (cell walls). Involved in grain production.

Manganese: Manganese is necessary for building the chloroplasts. Manganese deficiency may result in coloration abnormalities, such as discoloured spots on the foliage.

Sodium: Sodium is involved in the regeneration of phosphoenolpyruvate in CAM and C4 plants. It can also substitute for potassium in some circumstances.

Zinc: Zinc is required in a large number of enzymes and plays an essential role in DNA transcription. A typical symptom of zinc deficiency is the stunted growth of leaves, commonly known as "little leaf" and is caused by the oxidative degradation of the growth hormone auxin.

Nickel: In higher plants, Nickel is essential for activation of urease, an enzyme involved with nitrogen metabolism that is required to process urea. Without Nickel, toxic levels of urea accumulate, leading to the formation of necrotic lesions. In lower plants, Nickel activates several enzymes involved in a variety of processes, and can substitute for Zinc and Iron as a cofactor in some enzymes.

Cobalt has proven to be beneficial to at least some plants, but is essential in others, such as legumes where it is required for nitrogen fixation. Vanadium may be required by some plants, but at very low concentrations. It may also be substituting for molybdenum. Selenium and sodium may also be beneficial. Sodium can replace potassium's regulation of stomatal opening and closing.

Plant nutrition is a difficult subject to understand completely, partially because of the variation between different plants and even between different species or individuals of a given clone. Elements present at low levels may cause deficiency symptoms, and toxicity is possible at levels that are too high. Further, deficiency of one element may present as symptoms of toxicity from another element, and vice-versa. Carbon and oxygen are absorbed from the air, while other nutrients are absorbed from the soil. Green plants obtain their carbohydrate supply from the carbon dioxide in the air by the process of photosynthesis.

Integrated Nutrient Management

The main objective of Integrated Nutrient Management (INM) Division is to ensure adequate availability of quality fertilizers to farmers through periodical demand assessment and timely supply, promoting integrated nutrient management, which is soil test based, judicious and balanced use of chemical fertilizers in conjunction with organic manures and bio-fertilizers, promotion of organic farming and ensuring quality control of fertilizers through implementation of Fertilizer (Control) Order, 1985.

India is the third largest producer and consumer of fertilizers in the world after China and the USA. Against 18.4 million tonnes of fertilizer nutrients consumed during 2004-05, the nutrient consumption is estimated at 20.2 million tonnes during 2005-06. The consumption of major fertilizers namely, Urea, DAP, MOP, SSP and Complexes are estimated to be 22.2, 6.8, 2.6, 2.8 and 6.7 million tonnes respectively during 2005-06. India is, by and large, self sufficient in respect of Urea and about 90 per cent in case of DAP.

The all-India average fertilizer consumption is 96.4 kg./ha of NPK nutrients through there is wide variation from state to state varying from 197 kg/ha in Punjab, 164 kg/ha in Haryana to less than 10 kg/ha in States like Arunachal Pradesh, Nagaland, Sikkim, etc. Considering the skewed pattern of fertilizer use, Government of India is promoting balanced and integrated use of fertilizer nutrients through various initiatives.

As a result, NPK consumption ratio has now improved to 5.7: 2.2: 1 during 2004-05 from 7 : 2.7 : 1 during 2000-01.

Price of Fertilizers

Presently urea is the only fertilizer which is under Statutory Price Control. To ensure adequate availability of fertilizers to farmers at reasonable rates, subsidy is provided by Government of India. Urea, the most consumed fertilizer, is subsidised under the New Urea Pricing Scheme, whereas P&K fertilizers, which are decontrolled, are covered under the Concession Scheme. The existing scheme for special freight subsidy has been continued for supplies to the North Eastern States and Jammu and Kashmir. There is no change in the prices of major fertilizers level since 12 March 2003, which are as under:

Table: Prices of Major Fertilizers Level Since 12 March 2003

Fertilizer Product	*MRP (Rs. Per MT)*
Urea	4830
DAP (Indigenous)	9350
Complexes	7100-9080
MOP	4455
SSP	Price of SSP is fixed by State Governments

Buffer Stocking of P&K Fertilizers

To ensure adequate availability of decontrolled fertilizers in remote and inaccessible areas, there is a scheme under which a buffer stock of limited quantities of DAP and MOP is being maintained at strategic locations to meet emergent requirements.

Balanced Use of Fertilizers

The Government is implementing a Centrally Sponsored Scheme 'Balanced and Integrated Use of Fertilizers' to popularise soil test based judicious application of chemical fertilizers in combination with organic manures and bio-fertilizers. Under the Scheme, financial assistance is provided for setting up/strengthening of new soil testing laboratories for advising farmers on major and micro nutrients application, training and demonstration on balanced use of fertilizers and also for setting up of production units of useful organic manure from city waste/garbage. This scheme has since been subsumed within 'Macro Management Scheme' with effect from October 2000. State Governments are required to undertake all these activities through their Annual Work Plans.

Fertilizer Quality Control

The Government ensures the quality of fertilizers through Fertilizer Control Order (FCO) issued under Essential Commodities Act (ECA) to regulate the price, trade, quality and distribution of fertilizers in the country. The State Governments are the executing agencies to implement the various provisions of FCO. The Order strictly prohibits the manufacture, import and sale of any fertilizer, which does not meet prescribed standards.

The Central Fertilizer Quality Control and Training Institute at Faridabad and its 3 Regional centres located at Navi Mumbai, Chennai and Kalyani have been set up for inspection and analysis of imported and indigenous fertilizers, giving technical advice and providing training on quality control to state enforcement agencies and analysts. The Institute has also developed a Quick Testing Kit for on-the-spot detection of adulteration in fertilizers.

The FCO has been recently amended to make it more user friendly and ensuring effective enforcement. The important amendments include provision of repackaging of cut or torn fertilizer bags and reprocessing of damaged fertilizer due to natural calamities and provisions for commercial trials of provisional fertilizers. Accordingly, M/s National Fertilizer Limited (NFL), M/s Tata Chemicals, M/s Chambal Fertilizers Ltd., M/s Indo-Gulf Fertilizers and M/s Shri Ram Fertilizers have been permitted to manufacture neem-coated urea under Clause 20 A for commercial trials. M/s Coromandel Fertilizers Ltd. have been permitted to manufacture Bentonite Sulphur for commercial trials and M/s IFFCO has been permitted to manufacture fortified complex fertilizers. Printing of Maximum Retail Price (MRP) on all fertilizers including urea is now mandatory along with the printing of month and year of manufacture/import of fertilizer.

Presently there are 67 laboratories in the country (including 4 Central Government Laboratories) with a total annual analysing capacity of 1.25 lakh samples. The analytical capacity and the number of samples analysed and found non-standard during the last 5 years are as under:

Table: Analytical Capacity and the Number of Samples Analysed and found non-Standard during the last 5 Years

Year	*No. of Labs*	*Annual Analytical capacity*	*No. of samples analysed*	*Percentage of non-standard samples*
2000-01	66	120315	103436	5.3
2001-02	66	120415	104188	5.7
2002-03	67	125480	109504	5.4
2003-04	67	124778	104647	5.5
2004-05	67	124730	108859	6.0

The percentage of non-standard samples on all India basis is hovering around 5 -6 per cent samples during the last 5 years. However, there is a large variation from state to state as 18 per cent in M.P., 10-13 per cent in Maharashtra, H.P. and U.P., 4-8 per cent in Bihar, Orissa, W.B., Rajasthan, Haryana, Chhattisgarh, A.P., Kerala, Karnataka and Tamil Nadu and around 1-3 per cent in Punjab and Gujarat.

Promotion of Bio-fertilizers

With a view to promote the use of bio-fertilizers as an environment-friendly and cheaper source of plant nutrients, Government of India had earlier taken up a 'National Project on Development and Use of Bio-fertilizers' during the Sixth Plan and was continued till Ninth Plan. The Scheme provided for setting up National Bio-fertilizer Development Centre at Ghaziabad and its Six Regional Bio-fertilizer Development Centres (RBDCs) at Jabalpur, Nagpur, Bangalore, Bhubaneshwar, Hissar and Imphal for promotion, extension, training and demonstration of bio-fertilizers. Under the Scheme, assistance for setting up of bio-fertilizer production units was also provided.

Presently, there are around 125 bio-fertilizers production units in the country with an annual capacity of 18,000 tonnes of different types of bio-fertilizers and the annual production is about 10,000 tonnes. This scheme has been subsumed under a new Central Sector Scheme from October 2004.

To undertake the new activities, the names of NBDC and RBDCs have been changed to National Centre of Organic Farming (NCOF) and Regional Centres of Organic Farming (RCOF).

National Project on Organic Farming

The Government is implementing a new scheme 'National Project on Organic Farming' since October 2004 with an outlay of Rs 57.05 crore for production, promotion, market development of organic farming in the country during the remaining period of the Tenth Plan. The main components of the scheme include:

1. Putting in place a system of certification of organic produce.
2. Capacity building through service providers.
3. Financial support for commercial production units for production of organic inputs like:
    - Fruits and vegetable waste compost units;
    - Bio fertilizers production;
    - Hatcheries for vermi culture.
4. Promotion, extension and market development of Organic Farming.

Nutrient Management

Nutrient management is a process used by farmers to manage the amount, form, placement, and timing of the application of nutrients (whether as manure, commercial fertilizer, or other form of nutrients) to plants. The purpose is to supply plant nutrients for optimum forage and crop yields, to minimise non-point source pollution (run-off of pollutants to surface water) and contamination of groundwater, and to maintain and/or improve the condition of soil.

Nutrient Management Plan

A nutrient management plan is a set of conservation practices designed to use fertilizer and/or manure effectively while protecting against the potential adverse impacts of manure, erosion and organic byproducts on water quality. When such a plan is designed for animal feeding operations (AFO) it may be termed a "manure management plan." The plans typically address:

- soil testing;
- manure testing;
- erosion control practices;
- timing of fertilizer/manure application.

In the United States some regulatory agencies recommend or require that farms implement these plans, in order to prevent water pollution. The US Natural Resources Conservation Service (NRCS) has published guidance documents on preparing a comprehensive nutrient management plan (CNMP) for AFOs.

Integrated Nutrient Management

India is the third largest producer and consumer of fertilizers in the world after China and the USA, and contributes about 11.4 and 11.9 per cent to the total world production/consumption of NPK nutrients respectively. However, in terms of consumption per hectare, the consumption in India (104.5 kg per hectare) is lower than neighbouring countries, showing scope of further increase.

The consumption of chemical fertilizers in India by and large was stagnant from 2001-02 to 2003-04. However, it has increased during the last two years; consumption during 2005-06 was 203.40 lakh metric tonnes (LMT) of NPK nutrients which is the highest ever achieved. The consumption during the last few years is indicated below:

Consumption of fertilizers (in lakh tonnes)

Year	*Urea*	*DAP*	*MOP*	*N*	*P*	*K*	*N+P+K*	*Kg/ha*
1999-2000	202.78	69.37	20.49	115.92	47.99	16.78	180.69	94.94
2000-01	191.86	58.84	18.29	109.20	42.15	15.67	167.02	89.63
2001-02	199.17	61.81	19.93	113.10	43.82	16.67	173.60	91.13
2002-03	184.93	54.73	19.12	104.74	40.19	16.01	160.94	91.45
2003-04	197.67	56.24	18.41	110.77	41.24	15.98	167.99	88.05
2004-05	206.65	62.56	24.06	117.13	46.24	20.61	183.98	94.52
2005-06	222.97	67.64	27.31	127.23	52.04	24.13	203.40	104.50

The consumption of fertilizers during Kharif 2006 is estimated to be 100.10 LMT of NPK nutrients, which constitutes an increase of approximately 9 per cent over Kharif 2005. The consumption during 2006-07 (both Kharif and Rabi) is likely to be more than 220 LMT. However, the fertilizer consumption in India is very skewed with wide interstate, inter-district and inter-crop variations, though the NPK ratio, which is an indicator of balanced use of chemical fertilizers, has improved to 5.3:2.2:1 during 2005-06, against 5.7:2.2:1 during the preceding year.

To ensure the adequate availability of fertilizers to farmers, the Department of Agriculture and Cooperation makes a demand assessment well in advance through half-yearly input zonal conferences in consultation with the State Governments and the fertilizer industry. Thereafter, the Essential Commodities Act (ECA) Supply Plan and orders are issued under the Fertilizer Movement Control Order for indigenous and imported urea by the Department of Fertilizers to ensure its timely availability.

Fertilizer Prices: All chemical fertilizers except urea continued to be decontrolled. The Government of India continues to provide a subsidy to the manufacturers of urea to enable its availability to farmers at reasonable prices. Further, in order to make available decontrolled Phosphatic (P) and Potassic (K) fertilizers at reasonable prices, the government has been implementing a scheme of concession on sales of these fertilizers. There was no increase in the price of major fertilizers and the prices of urea and other fertilizers remained at the same level as of 2003-04. The fertilizers during the year under report are as under:

S.No.	Product	Rs per tonne
1.	Urea	4830
2.	Di-Ammonium Phosphate (DAP)	9350
3.	Muriate of Potash (MoP)	4455
4.	Single Super Phosphate (SSP)	Prices fixed by the State Governments vary from 2600-4612
5.	Complex Fertilizers	6980-9080

Buffering Stocking of P and K Fertilizers: To ensure the adequate availability of decontrolled fertilizers in remote and inaccessible areas, a buffer stock of limited quantities of DAP and MOP is being maintained at strategic locations to meet the emergent requirements of states.

Promotion of Integrated Nutrient Management (INM): Some major initiatives that have been taken to promote the balanced and integrated use of fertilizers are enumerated below:

The Government is promoting the soil test-based balanced and judicious use of chemical fertilizers, bio-fertilizers and locally available organic manures like farmyard manure, compost, nadep compost, vermi compost, green manure and press mud, etc., to maintain soil health and its productivity. The Centrally Sponsored Scheme on Balanced and Integrated Use of Fertilizers, since subsumed under the Macro Management of Agriculture Scheme, provides for the promotion of soil test-based application of chemical fertilizers, strengthening of soil testing facilities in the country and setting up of compost plants for conversion of biodegradable city waste into organic manure.

At present, there are 609 soil-testing laboratories in India. These include 487 static and 122 mobile laboratories under the State Governments and the fertilizer industry with an annual analysing capacity of 6.7 million soil samples. Under the scheme, soil health cards are being issued by the State Governments to the farmers for advising them on the use of correct and balanced use of fertilizers for maximum efficiency and profitability.

Status of Fertilizer Quality Control: To ensure the adequate availability of fertilizers of a standard quality to farmers, fertilizers were declared as an essential commodity; and the Fertilizer Control Order (FCO), 1985, was promulgated under Section 3 of the ECA to regulate the trade, price, quality and distribution of fertilizers in India. The FCO has been recently amended to make it more user-friendly and amenable to effective enforcement. For the first time, bio-fertilizers and organic manures have also been brought under the regulatory mechanism. In Schedule III and IV of the FCO the specifications of four important bio-fertilizers and three organic manures have been notified to ensure the adequate availability of fertilizers of standard quality to farmers.

The enforcement of this order has primarily been entrusted to State Governments. The Central Government provides training facilities and technical guidance to states and supplements their efforts through random inspection of manufacturing units and their distribution network through inspectors. At present, there are 67 laboratories in India (including four Central Government laboratories), with a total annual fertilizer capacity of 1.22 lakh samples. The analytical capacity and the number of samples analysed and found to be non-standard during the last five years are as under:

Year	*No. of labs*	*Annual analytical capacity*	*No. of samples analysed*	*% Non-standard samples*
2001-02	65	119415	108425	5.8
2002-03	67	125480	109504	5.4
2003-04	67	124778	104647	5.5
2004-05	67	124730	108859	6.0
2005-06	67	122488	111745	6.0
2006-07 (up to 30 September 2006)	67	122500	31749*	4.3

*Information awaited from some states.

The Central Fertilizer Quality Control & Training Institute (CFQC&TI or the Institute), Faridabad: The major functions of the CFQC&TI and its three Regional Fertilizer Control Laboratories (RFCLs) located at Navi Mumbai, Chennai and Kalyani include drawal, inspection and analysis of both indigenous and imported fertilizers; training of state enforcement officers and analysts including officers from developing countries; development of methods of analysis; and acting as a referee laboratory besides an advisory body on the issues relating to fertilizer quality control. The activities undertaken during the year under report are briefly indicated below:

The CFQC&TI and RFCLs analysed 14189 samples against the target of 8500 during 2005-06 (showing an achievement of 166.9 per cent over the target), out of which 402 samples were found to be non-standard. During 2006-07 (till 31 December 2006), a further 11004 samples were analysed, out of which 261 samples were found to be non-standard.

During 2005-06, 517 ships were inspected, out of which seven containers of 100 per cent water soluble NPK and one urea ship were found to be non-standard. The Institute/RFCLs' inspectors inspected 490 ships and containers, of which six containers of NPK (100 per cent water soluble) were found to be non-standard in nutrient form during 2006-07 (up to 31 December 2006).

The central teams, during a random inspection of manufacturing units and dealers' premises, collected 120 samples during 2005-06, of which 29.2 per cent samples were found to be non-standard. During 2006-07 (up to 31 December 2006), 87 samples were drawn, of which 35 per cent samples were found to be non-standard.

During 2006-07 (up to 31 December 2006), 38 training programmes for fertilizer inspectors/analysts were organised. In addition, one exclusive training programme for foreign participants (in which 20 officers from 15 countries participated) was held.

Training Programme for North-Eastern States: One refresher training course of two weeks' duration was organised in September 2006, exclusively for the benefit of enforcement officers of the North-Eastern states at RFCL, Kalyani. In addition, one two-day orientation training programme was organised in Assam in December 2006; and two dealers' programmes were also organised.

National Seminar on Fertilizer Quality Control: The Institute organised a National Seminar on Fertilizer Quality Control on the 11 and 12 April 2005, at Faridabad. About 250 participants, mainly from the State Governments, ICAR, SAUs, the fertilizer industry, importers/surveyors and fertilizer testing laboratories, attended the seminar. After in-depth discussions, the plenary session unanimously adopted 20 recommendations on fertilizer quality control. Most of the recommendations have since been accepted by the Central Government.

National Project on Organic Farming: This scheme was initiated as a pilot project in October 2004 for the production, promotion, certification and market development of organic farming in India, with an outlay of Rs 57.05 crore. The project was introduced by subsuming the existing infrastructure of the National Bio-fertilizer Development Centre, Ghaziabad, and its six regional centres, located at Hissar, Jabalpur, Bangalore, Nagpur, Bhubaneshwar and Imphal.

During the period 2006-07 (up to 31 December 2006), funds were provided for conducting 78 programmes for capacity building; 10 programmes on certification and inspection; 27 programmes on production and quality control of organic inputs; 79 programmes for the benefit of field functionaries (extension officers; and 219 training courses for farmers. In addition, as a part of the project, 68 model organic farms; 1093 field demonstrations on organic inputs; and 145 field demonstrations on enriched biogas slurry were sanctioned. Also, during the year under report (till 31 October 2006), 1014 samples of different organic inputs were analysed of which 23 of the samples were found to be substandard.

3

Fertilizers

Fertilizers are chemical compounds applied to promote plant and fruit growth. Fertilizers are usually applied either through the soil (for uptake by plant roots) or, by foliar feeding (for uptake through leaves).

Fertilizers can be placed into the categories of *organic fertilizers* (composed of decayed plant/animal matter), or *inorganic fertilizers* (composed of simple chemicals and minerals). *Organic fertilizers* are 'naturally' occurring compounds, such as peat, manufactured through natural processes (such as composting), or naturally occurring mineral deposits; *inorganic fertilizers* are manufactured through chemical processes (such as the Haber process), also using naturally occurring deposits, while chemically altering them (e.g. concentrated triple superphosphate).

Properly applied, organic fertilizers can improve the health and productivity of soil and plants, as they provide different essential nutrients to encourage plant growth. Organic nutrients increase the abundance of soil organisms by providing organic matter and micronutrients for organisms such as fungal mycorrhiza, which aid plants in absorbing nutrients. Chemical fertilizers may have long-term adverse impact on the organisms living in soil and a detrimental long term effect on soil productivity of the soil.

Chemical Content

Fertilizers typically provide, in varying proportions, the three major plant nutrients: nitrogen, phosphorus, potassium known shorthand as N-P-K); the secondary plant nutrients (calcium, sulphur, magnesium) and sometimes trace elements (or micronutrients) with a role in plant or animal nutrition: boron, chlorine, manganese, iron, zinc, copper, molybdenum and (in some countries) selenium.

Organic and Non-organic Fertilizers

Both organic and inorganic fertilizers were called "manure", derived from the French expression for *manual* (of or belonging to the hand) tillage, however, this term is currently restricted to organic manure. Though nitrogen is plentiful in the Earth's atmosphere, relatively few plants engage in nitrogen fixation (conversion of atmospheric nitrogen to a plant-accessible form).

It is believed by some that 'organic' agricultural methods are more environmentally friendly and better maintain soil organic matter (SOM) levels. There are some scientific studies that support this position.

Historical Backdrop

While manure, cinder and iron-making slag have been used to improve crops for centuries, the use of fertilizers is arguably one of the great innovations of the Agricultural Revolution of the 19th Century.

Key Figures in Europe: In the 1730s, Viscount Charles Townshend (1674-1738) first studied the improving effects of the four crop rotation system that he had observed in use in Flanders. For this he gained the nickname of Turnip Townshend.

Justus von Liebig: Chemist Justus von Liebig (1803-1883) contributed greatly to the advancement in the understanding of plant nutrition. His influential works first denounced the vitalist theory of humus, arguing first the importance of ammonia, and later promoting the importance of inorganic minerals to plant nutrition. Primarily Liebig's work succeeded in exposition of questions for agricultural science to address over the next 50 years.

In England, he attempted to implement his theories commercially through a fertilizer created by treating phosphate of lime in bone meal with sulphuric acid. Although it was much less expensive than the guano that was used at the time, it failed because it was not able to be properly absorbed by crops.

Sir John Bennet Lawes: At that time in England, Sir John Bennet Lawes (1814-1900) was experimenting with crops and manures at his farm at Harpenden and was able to produce a practical superphosphate in 1842 from the phosphates in rock and coprolites. Encouraged, he employed Sir Joseph Henry Gilbert, who had studied under Liebig at the University of Giessen, as director of research. To this day, the Rothamsted research station the pair founded still investigates the impact of inorganic and organic fertilizers on crop yields.

Jean Baptiste Boussingault: In France, Jean Baptiste Boussingault (1802-1887) pointed out that the amount of nitrogen in various kinds of fertilizers is important.

Metallurgists Percy Gilchrist (1851-1935) and Sidney Gilchrist Thomas (1850-1885) invented the Thomas-Gilchrist converter, which enabled the use of high phosphorus acidic Continental ores for steel-making. The dolomite lime lining of the converter turned in time into calcium phosphate, which could be used as fertilizer, known as Thomas-phosphate.

Bosch Farben and Haber: In the early decades of the 20th Century, the Nobel Prize winning chemists Carl Bosch of I.G. Farben and Fritz Haber developed the process that enabled nitrogen to be synthesised cheaply into ammonia, for subsequent oxidation into nitrates and nitrites.

Erling Johnson: In 1927 Erling Johnson developed an industrial method for producing nitrophosphate, also known as the Odda process after his Odda Smelteverk of Norway. The process involved acidifying phosphate rock (from Nauru and Banaba Islands in the southern Pacific Ocean) with nitric acid to produce phosphoric acid and calcium nitrate which, once neutralised, could be used as a nitrogen fertilizer.

Inorganic Fertilizers

Naturally occurring inorganic fertilizers include Chilean sodium nitrate, mined rock phosphate, and limestone (to raise pH and a calcium source).

Macronutrients and Micronutrients

Fertilizers can be divided into macronutrients and micronutrients based on their concentrations in plant dry matter. There are six macronutrients: nitrogen, phosphorus, and potassium, often termed "primary macronutrients" because their availability is usually managed with NPK fertilizers, and the "secondary macronutrients" – calcium, magnesium, and sulphur – which are required in roughly similar quantities but whose availability is often managed as part of liming and manuring practices rather than fertilizers.

The macronutrients are consumed in larger quantities and normally present as a whole number or tenths of percentages in plant tissues (on a dry matter weight basis). There are many micronutrients, required in concentrations ranging from 5 to 100 parts per million (ppm) by mass. Plant micronutrients include iron (Fe), manganese (Mn), boron (B), copper (Cu), molybdenum (Mo), nickel (Ni), chlorine (Cl), and zinc (Zn).

Macronutrient Fertilizers

Synthesised materials are also called artificial, and may be described as straight, where the product predominantly contains the three primary ingredients of nitrogen (N), phosphorus (P), and potassium (K), (known as N-P-K fertilizers or compound fertilizers when elements are mixed intentionally).

Reporting of N-P-K: Such fertilizers are named according to the content of these three elements. For example, if nitrogen is the main element, the fertilizer is often described as a nitrogen fertilizer.

Regardless of the name, however, they are labelled according to the relative amounts of each of these three elements, by weight (i.e., mass fraction). The percent of nitrogen is reported directly. However, phosphorus is reported as the mass fraction of phosphorus pentoxide (P_2O_5), the anhydride of phosphoric acid, and potassium is reported as the mass fraction of potassium oxide (K_2O), which is the anhydride of potassium hydroxide.

Fertilizer composition is expressed in this fashion for historical reasons in the way it was analysed (conversion to ash for P and K mass fractions); this practice dates back to Justus von Liebig.

Mass Fraction Conversion to Elemental Values

Since the N-P-K reporting basis just described does not give the actual fraction of the respective elements, some packaging also reports the elemental mass fractions. The UK fertilizer-labelling regulations allow for additionally reporting the elemental mass fractions of phosphorous and potassium, rather than phosphoric acid and potassium hydroxide, but these must be listed in parentheses after the standard values. The regulations specify the factors for converting from the P_2O_5 and K_2O values to the respective P and K elemental values as follows:

In phosphorous pentoxide, the element phosphorous constitutes 43.6% of the total mass of the compound. Thus, the official UK mass fraction (percentage) of elemental phosphorus is 43.6%. $[P] = 0.436 \times [P_2O_5]$

Likewise, the mass fraction (percentage) of elemental potassium is 83%. $[K] = 0.83 \times [K_2O]$

Thus an 18"51"20 fertilizer contains, by weight, 18% elemental nitrogen (N), 22% elemental phosphorus (P), and 16% elemental potassium (K).

(Note: The remaining 11% [100 - (18 + 51 + 20)] is known as ballast or filler and may or may not be valuable to the plants, depending on what is used as filler.)

Nitrogen Fertilizer

Nitrogen fertilizer is often synthesised using the Haber-Bosch process, which produces ammonia. This ammonia can be applied directly to the soil, or used to produce other compounds (notably anhydrous ammonium nitrate and urea). These concentrated products may be used as fertilizer or diluted with water to form a concentrated liquid fertilizer, UAN. Ammonia can also be used in the Odda Process in combination with rock phosphate and potassium fertilizer to produce compound fertilizers.

Table: Major users of Nitrogen-based Fertilizer

Country	*Total N consumption (Mt pa)*	*of which used for feed & pasture*
USA	9.1	4.7
China	18.7	3.0
France	2.5	1.3
Germany	2.0	1.2
Canada	1.6	0.9
UK	1.3	0.9
Brazil	1.7	0.7
Spain	1.2	0.5
Mexico	1.3	0.3
Turkey	1.5	0.3
Argentina	0.4	0.1

The production of ammonia currently consumes about 5% of global natural gas consumption, which is somewhat under 2% of world energy production.

Natural gas is overwhelmingly used for the production of ammonia, but other energy sources, together with a hydrogen source, can be used for the production of nitrogen compounds suitable for fertilizers. The cost of natural gas makes up about 90% of the cost of producing ammonia. The price increases in natural gas in the past decade, along with other factors such as increasing demand, have contributed to an increase in fertilizer price.

Nitrogen-based fertilizers are most commonly used to treat fields used for growing maize, followed by barley, sorghum, rapeseed, soyabean and sunflower. One study has shown that application of nitrogen fertilizer on off-season cover crops can increase the biomass of these crops, while having a beneficial effect on soil nitrogen levels for the cash crop planted during the summer season.

Agricultural versus Horticultural Fertilizers

In general, agricultural fertilizers contain only 1 or 2 macronutrients. Agricultural fertilizers are intended to be applied infrequently and normally prior to or alongside seeding. Examples of agricultural fertilizers are granular triple superphosphate, potassium chloride, urea, and anhydrous ammonia. The commodity nature of fertilizer, combined with the high cost of shipping, may lead to use of locally available substitutes or materials from the closest and/or cheapest source, which may vary with factors such as the relative cost of transportation by rail, ship, or truck.

In other words, a particular nitrogen source may be very popular in one part of the country while another is very popular in another geographic region only due to factors unrelated to agronomic concerns.

Horticultural fertilizers, on the other hand, are formulated from many of the same compounds and some others to produce well-balanced fertilizers that also contain micronutrients. Some materials, such as ammonium nitrate, are used minimally in large scale production farming. The 18-51-20 example is a horticultural fertilizer formulated with high phosphorus to promote bloom development in ornamental flowers. Horticultural fertilizers may be water-soluble (instant-release) or relatively insoluble (controlled-release).

Controlled release fertilizers are also referred to as sustained-release or timed-release. Many controlled release fertilizers are intended to be applied approximately every 3–6 months, depending on watering, growth rates, and other conditions, whereas water-soluble fertilizers must be applied at least every 1–2 weeks and can be applied as often as every watering if sufficiently dilute.

Unlike agricultural fertilizers, horticultural fertilizers are marketed directly to consumers and become part of retail product distribution lines.

Health and Sustainability Issues

Many inorganic fertilizers do not replace *trace* mineral elements in the soil which become gradually depleted by crops. This depletion has been linked to studies which have shown a marked fall (up to 75%) in the quantities of such minerals present in fruit and vegetables. One exception is in Western Australia where deficiencies of zinc, copper, manganese, iron and molybdenum were identified as limiting the growth of crops and pastures in the 1940s and 1950s. Soils in Western Australia are very old, highly weathered and deficient in many of the major nutrients and trace elements. Since this time these trace elements are routinely added to inorganic fertilizers used in agriculture in this state.

In many countries there is the public perception that inorganic fertilizers "poison the soil" and result in "low quality" produce. However, there is very little (if any) scientific evidence to support these views. When used appropriately, inorganic fertilizers enhance plant growth, the accumulation of organic matter, and the biological activity of the soil, thus preventing overgrazing and soil erosion. The nutritional value of plants for human and animal consumption is typically improved when inorganic fertilizers are used appropriately.

There are concerns regarding arsenic, cadmium and uranium accumulating in fields treated with fertilizers. The phosphate minerals contain trace amounts of these elements and if no cleaning step is applied after mining the continuous use of phosphate fertilizers leads towards an accumulation of these elements in the soil.

Phosphate fertilizers replace inorganic arsenic naturally found in the soil, displacing the heavy metal and causing accumulation in run-off. Eventually these heavy metals can build up to unacceptable levels and build up in produce.

Another problem with *inorganic* fertilizers is that they are now produced in ways which cannot be continued indefinitely. Potassium and phosphorus come from mines (or saline lakes such as the Dead Sea) and such resources are limited. Nitrogen sources are effectively unlimited (forming over 70% of atmospheric gases), however, nitrogen fertilizers are presently made using fossil fuels such as natural gas and coal, which are limited.

Innovative thermal depolymerisation biofuel schemes are experimenting with the production of by-products with 9% nitrogen fertilizer from organic waste.

Organic Fertilizers

Naturally occurring organic fertilizers include manure, worm castings, peat moss, seaweed, sewage and guano. Sewage sludge use in organic agricultural operations in the US has been extremely limited and rare due to USDA prohibition of the practice (due to toxic metal accumulation, among other factors).

Cover crops are also grown to enrich soil as a green manure through nitrogen fixation from the atmosphere by bacterial nodules on roots; as well as phosphorus (through nutrient mobilisation) content of soils.

Processed organic fertilizers from *natural* sources include compost (from green waste), bloodmeal and bone meal (from organic meat production facilities), and seaweed extracts (alginates and others).

Mixed Definitions of Organic Fertilizer

There can be confusion as to the veracity of the term 'organic' when applied to agricultural systems and fertilizer.

The problem is one of confusion of terminology between agricultural and chemical disciplines.

Minerals such as mined rock phosphate, sulphate of potash and limestone are also considered *organic fertilizers,* although they contain no organic (carbon) molecules. Some ambiguity in the usage of the term *organic* exists; however, it is simple to differentiate with a separation between the scientific and colloqial uses [as in *velocity* in common usage (speed) and physics usage (velocity)].

Synthetic fertilizers, such as urea and urea formaldehyde, are organic in the sense of the organic chemistry definition of organic, can be supplied *organically* (agriculturally), but when manufactured as a pure chemical is not *organic* under organic certification standards.

Naturally mined powdered limestone, mined rock phosphate and sodium nitrate, are *inorganic* (in a chemical sense) in that they contain no carbon molecules, and are energetically-intensive to harvest, *but* are approved for organic agriculture in *minimal* amounts.

The common thread that can be seen through these examples is that *organic* agriculture defines itself through minimal processing (e.g. via chemical energy such as petroleum), as well as being naturally-occurring (as is, or via natural biological processing such as the composting process).

Benefits of Organic Fertilizers

However, by their nature, organic fertilizers provide increased physical and biological storage mechanisms to soils, mitigating risks of over-fertilization. Organic fertilizer nutrient content, solubility, and nutrient release rates are typically much lower than mineral (inorganic) fertilizers. One study found that over a 140-day period, after 7 leachings:

- Organic fertilizers had released between 25% and 60% of their nitrogen content;
- Controlled Release Fertilizers (CRFs) had a relatively constant rate of release;

- Soluble fertilizer released most of its nitrogen content at the first leaching.

Disadvantages of Organic Fertilizers

It is difficult to chemically distinguish between urea of biological origin and those produced synthetically. Like chemical fertilizers, it is possible to over-apply organic fertilizers if does not measure and distribute the required amounts according to the recommended amounts for the plot of land in question.. Release of the nutrients may happen quite suddenly depending on the type of organic fertilizer used.

Environmental Risks of Fertilizer Use

High application rates of *inorganic* nitrogen fertilizers in order to maximise crop yields, combined with the high solubilities of these fertilizers leads to increased leaching of nitrates into groundwater.

Eventually, nitrate-enriched groundwater makes its way into lakes, bays and oceans where it accelerates the growth of algae, disrupts the normal functioning of water ecosystems, and kills fish in a process called eutrophication (which may cause water to become cloudy and/or discoloured—green, yellow, brown, or red). About half of all the lakes in the United States are now eutrophic, while the number of oceanic dead zones near inhabited coastlines are increasing.

The use of ammonium nitrate in *inorganic* fertilizers is particularly damaging, as plants absorb ammonium ions preferentially over nitrate ions. This allows excess nitrate ions which are not absorbed to be freely dissolved (by rain or irrigation) into groundwater and other waterways, leading to eutrophication.

Nitrate levels above 10 mg/L (10 ppm) in groundwater can cause 'blue baby syndrome' (acquired methemoglobinemia), leading to hypoxia (which can lead to coma and death if not treated).

As of 2006, the application of nitrogen fertilizer is being increasingly controlled in Britain and the United States. If eutrophication *can* be reversed, it may take decades before the accumulated nitrates in groundwater can be broken down by natural processes.

Storage and application of some nitrogen fertilizers in some weather or soil conditions can cause emissions of the greenhouse gas nitrous oxide (N_2O). Ammonia gas (NH_3) may be emitted following application of 'inorganic' fertilizers, or manure/slurry. Besides supplying nitrogen, ammonia can also increase soil acidity (lower pH, or "souring"). Excessive nitrogen fertilizer applications can also lead to pest problems by increasing the birth rate, longevity and overall fitness of certain pests.

The concentration of up to 100 mg/kg of cadmium in phosphate minerals (for example, minerals from Nauru and the Christmas islands) increases the contamination of soil with cadmium, for example in New Zealand. Uranium is another example of a contaminant often found in phosphate fertilizers; also, radioactive Polonium-210 contained in phosphate fertilizers is absorbed by the roots of plants and stored in its tissues. Tobacco derived from plants fertilzed by rock phosphates contains Polonium-210 which emits alpha radiation estimated to cause about 11,700 lung cancer deaths each year worldwide.

For these reasons, it is recommended that knowledge of the nutrient content of the soil and nutrient requirements of the crop are carefully balanced with application of nutrients in inorganic fertilizer. This process is called nutrient budgeting. By careful monitoring of soil conditions, farmers can avoid wasting expensive fertilizers, and also avoid the potential costs of cleaning up any pollution created as a by-product of their farming.

Hazard of Over-fertilization

Over-fertilization of a vital nutrient can be as detrimental as underfertilization. "Fertilizer burn" can occur when too much fertilizer is applied, resulting in a drying out of the roots and damage or even death of the plant.

According to UC IPM, all *organic fertilizers,* and some specially-formulated *inorganic fertilizers* are classified as 'slow-release' fertilizers, and therefore cannot cause nitrogen burn Organic fertilizers are as likely to cause plant burn as inorganic fertilizers.

If excess nitrogen is present, some plants can exude the excess through their leaves in a process called guttation.

Environmental Toxicity of Fertilizers

Toxic fertilizers are recycled industrial waste that introduce several classes of toxic materials into farm land, garden soils, and water streams. The consumption levels of toxic fertilizer are increasing lately in the US from citizens who are purchasing the wrong chemicals for their gardens as well as choosing the wrong company to purchase it from.

This is leading to major environmental problems due to the fact of toxic waste being processed and planted into our land and water. The most common toxic elements in this type of fertilizer are mercury, lead, and arsenic.

Between 1990-1995, 600 companies from 44 different states sent 270 million pounds of toxic waste to farms and fertilizer companies across the country.

However, according to the United States Food and Drug Administration:

> "Current information indicates that only a relatively small percentage of fertilizers is manufactured using industrial wastes as ingredients, and that hazardous wastes are used as ingredients in only a small portion of waste-derived fertilizers."

and

> "the EPA has continually encouraged the beneficial reuse and recycling of industrial wastes."

Heavy Metal Content of Recycled Fertilizer

Steel industry wastes, recycled into fertilizers for their high levels of zinc (essential to plant growth), wastes can include the following toxic metals:

- lead;
- arsenic;
- cadmium;
- chromium and
- nickel.

Toxic Organic Compounds

Dioxins, polychlorinated dibenzo-p-dioxins (PCDDs), and polychlorinated dibenzofurans (PCDFs) have been detected in fertilizers and soil amendments.

Global Issues

"We throw away nutrients for our plants in underground sewage systems. We do this in such a way that pollutes underground water tables. Then we buy manufactured "nutrients" for our plants which aren't as good as what we threw away. This is modern day wastewater "technology". Michael Reynolds — Earthship Vol.2: Systems and Components".

The growth of the world's population to its current figure has only been possible through intensification of agriculture associated with the use of fertilizers. There is an impact on the sustainable consumption of other global resources as a consequence.

The use of fertilizers on a global scale emits significant quantities of greenhouse gas into the atmosphere. Emissions come about through the use of:

- animal manures and urea, which release methane, nitrous oxide, ammonia, and carbon dioxide in varying quantities depending on their form (solid or liquid) and management (collection, storage, spreading);

- fertilizers that use nitric acid or ammonium bicarbonate, the production and application of which results in emissions of nitrogen oxides, nitrous oxide, ammonia and carbon dioxide into the atmosphere.

By changing processes and procedures, it is possible to mitigate some, but not all, of these effects on anthropogenic climate change.

The nitrogen-rich compounds found in fertilizer run-off is the primary cause of a serious depletion of oxygen in many parts of the ocean, especially in coastal zones; the resulting lack of dissolved oxygen is greatly reducing the ability of these areas to sustain oceanic fauna.

Soil Conditioner

A soil conditioner, also called a soil amendment, is a material added to soil to improve plant growth and health. The type of conditioner added depends on the current soil composition, climate, and the type of plant. Some soils lack nutrients necessary for proper plant growth and others hold too much or too little water. A conditioner or a combination of conditioners corrects the soil's deficiencies. Lime is used to make soil less acidic, as is lime-containing crushed stone. Fertilizers, such as peat, manure, anaerobic digestate or compost add depleted plant nutrients. Materials such as diatomaceous earth, clay, vermiculite, hydrogel, and shredded bark will make soil hold more water. Gypsum releases nutrients and improves soil structure. Sometimes a soil inoculant is added for legumes.

Unless clay is incorporated into a healthy crumb structure, water may bond to it too strongly to be available to plant roots.

Solubility

Solubility is the property of a solid, liquid, or gaseous chemical substance called solute to dissolve in a liquid solvent to form a homogeneous solution. The solubility of a substance strongly depends on the used solvent as well as on temperature and pressure. The extent of the solubility of a substance in a specific solvent is measured as the saturation concentration where adding more solute does not increase the concentration of the solution.

The solvent is generally a liquid, which can be a pure substance or a mixture. One also speaks of solid solution, but rarely of solution in a gas.

The extent of solubility ranges widely, from infinitely soluble (fully miscible) such as ethanol in water, to poorly soluble, such as silver chloride in water. The term *insoluble* is often applied to poorly or very poorly soluble compounds.

Under certain conditions the equilibrium solubility can be exceeded to give a so-called supersaturated solution, which is metastable.

Molecular View

Solubility occurs under dynamic equilibrium, which means that solubility results from the simultaneous and opposing processes of dissolution and phase separation (e.g. precipitation of solids]]). The solubility equilibrium occurs when the two processes proceed at a constant rate.

The term *solubility* is also used in some fields where the solute is altered by solvolysis. For example, many metals and their oxides are said to be "soluble in hydrochloric acid," whereas the aqueous acid degrades the solid to irreversibly give soluble products. It is also true that most ionic solids are degraded by polar solvents, but such processes are reversible. In those cases where the solute is not recovered upon evaporation of the solvent, the process is referred to as solvolysis. The thermodynamic concept of solubility does not apply straightforwardly to solvolysis.

When a solute dissolves, it may form several species in the solution. For example, an aqueous suspension of ferrous hydroxide, $Fe(OH)_2$, will contain the series $[Fe(H_2O)_{6-x}(OH)_x]^{(2-x)+}$ as well as other oligomeric species. Furthermore, the solubility of ferrous hydroxide and the composition of its soluble components depends on pH. In general, solubility in the solvent phase can be given only for a specific solute which is thermodynamically stable, and the value of the solubility will include all the species in the solution.

Factors Affecting Solubility

Solubility is defined for specific phases. For example, the solubility of aragonite and calcite in water are expected to differ, even though they are both polymorphs of calcium carbonate and have the same chemical formula.

The solubility of one substance in another is determined by the balance of intermolecular forces between the solvent and solute, and the entropy change that accompanies the solvation. Factors such as temperature and pressure will alter this balance, thus changing the solubility.

Solubility may also strongly depend on the presence of other species dissolved in the solvent, for example, complex-forming anions (ligands) in liquids. Solubility will also depend on the excess or deficiency of a common ion in the solution, a phenomenon known as the common-ion effect. To a lesser extent, solubility will depend on the ionic strength of solutions. The last two effects can be quantified using the equation for solubility equilibrium.

For a solid that dissolves in a redox reaction, solubility is expected to depend on the potential (within the range of potentials under which the solid remains the thermodynamically stable phase). For example, solubility of gold in high-temperature water is observed to be almost an order of magnitude higher when the redox potential is controlled using a highly-oxidising Fe_3O_4-Fe_2O_3 redox buffer than with a moderately-oxidising Ni-NiO buffer.

Solubility (metastable) also depends on the physical size of the crystal or droplet of solute (or, strictly speaking, on the specific or molar surface area of the solute). For highly defective crystals, solubility may increase with the increasing degree of disorder. Both of these effects occur because of the dependence of solubility constant on the Gibbs energy of the crystal. The last two effects, although often difficult to measure, are of practical importance. For example, they provide the driving force for precipitate aging (the crystal size spontaneously increasing with time).

Temperature

The solubility of a given solute in a given solvent typically depends on temperature. For many solids dissolved in liquid water, the solubility increases with temperature up to 100 °C. In liquid water at high temperatures, (e.g., that approaching the critical temperature), the solubility of ionic solutes tends to decrease due to the change of properties and structure of liquid water; the lower dielectric constant results in a less polar solvent.

Gaseous solutes exhibit more complex behaviour with temperature. As the temperature is raised, gases usually become less soluble in water, but more soluble in organic solvents.

Many salts behave like barium nitrate and disodium hydrogen arsenate, and show a large increase in solubility with temperature. Some solutes (e.g. NaCl in water) exhibit solubility which is fairly independent of temperature. A few, such as cerium (III) sulphate, become less soluble in water as temperature increases. This temperature dependence is sometimes referred to as "retrograde" or "inverse" solubility. Occasionally, a more complex pattern is observed, as with sodium sulphate, where the less soluble decahydrate crystal loses water of crystallisation at 32 °C to form a more soluble anhydrous phase.

The solubility of organic compounds nearly always increases with temperature. The technique of recrystallisation, used for purification of solids, depends on a solute's different solubilities in hot and cold solvent. A few exceptions exist, such as certain cyclodextrins.

Pressure

For condensed phases (solids and liquids), the pressure dependence of solubility is typically weak and usually neglected in practice. Assuming an ideal solution, the dependence can be quantified as:

$$\left(\frac{\partial \ln N_i}{\partial P}\right)_T = -\frac{V_{i,aq} - V_{i,cr}}{RT}$$

where the index i iterates the components, N_i is the mole fraction of the *i*th component in the solution, *P* is the pressure, the index T refers to constant temperature, $V_{i,aq}$ is the partial molar volume of the *i*th component in the solution, $V_{i,cr}$ is the partial molar volume of the *i*th component in the dissolving solid, and *R* is the universal gas constant.

The pressure dependence of solubility does occasionally have practical significance. For example, precipitation fouling of oil fields and wells by calcium sulphate (which decreases its solubility with decreasing pressure) can result in decreased productivity with time.

Solubility of Gases

Henry's law is used to quantify the solubility of gases in solvents. The solubility of a gas in a solvent is directly proportional to the partial pressure of that gas above the solvent. This relationship is written as:

$$pk = kc$$

where,

k is a temperature-dependent constant (for example, 769.2 L atm/mol for dioxygen (O_2) in water at 298 K),

p is the partial pressure (atm), and

c is the concentration of the dissolved gas in the liquid (mol/L).

Polarity

A popular aphorism used for predicting solubility is "*like dissolves like*". This statement indicates that a solute will dissolve best in a solvent that has a similar polarity to itself. This view is rather simplistic, since it ignores many solvent-solute interactions, but it is a useful rule of thumb. For example, a very polar (hydrophilic) solute such as urea is very soluble in highly polar water, less soluble in fairly polar methanol, and practically insoluble in non-polar solvents such as benzene. In contrast, a non-polar or lipophilic solute such as naphthalene is insoluble in water, fairly soluble in methanol, and highly soluble in non-polar benzene.

Liquid solubilities also generally follow this rule. Lipophilic plant oils, such as olive oil and palm oil, dissolve in non-polar solvents such as alkanes, but are less soluble in polar liquids such as water.

Synthetic chemists often exploit differences in solubilities to separate and purify compounds from reaction mixtures, using the technique of liquid-liquid extraction.

Insolubility and spontaneous phase separation does not mean that dissolution is disfavoured by enthalpy. Quite the contrary, in the case of water and hydrophobic substances, *hydrophobic hydration* is reasonably exothermic and enthalpy alone should be favour it. It appears that entropic factors – the reduced freedom of movement of water molecules around hydrophobic molecules – lead to an overall hydrophobic effect.

Rate of Dissolution

Dissolution is not always an instantaneous process. It is fast when salt and sugar dissolve in water but much slower for a tablet of aspirin or a large crystal of hydrated copper (II) sulphate. These observations are the consequence of two factors: the rate of solubilisation is related to the solubility product and the surface area of the material.

The speed at which a solid dissolves may depend on its crystallinity or lack thereof in the case of amorphous solids and the surface area (crystallite size) and the presence of polymorphism. Many practical systems illustrate this effect, for example in designing methods for controlled drug delivery.

Critically, the dissolution rate depends on the presence of mixing and other factors that determine the degree of undersaturation in the liquid solvent film immediately adjacent to the solid solute crystal.

In some cases, solubility equilibria can take a long time to establish (hours, days, months, or many years; depending on the nature of the solute and other factors). In practice, it means that the amount of solute in a solution is not always determined by its thermodynamic solubility, but may depend on kinetics of dissolution (or precipitation).

The rate of dissolution and solubility should not be confused as they are different concepts, kinetic and thermodynamic, respectively. The solubilisation kinetics, as well as apparent solubility can be improved after complexation of an active ingredient with cyclodextrin. This can be used in the case of drug with poor solubility.

Quantification of Solubility

Solubility is commonly expressed as a concentration, either by mass (g of solute per kg of solvent, g per dl (100 ml) of solvent), molarity, molality, mole fraction or other similar descriptions of concentration. The maximum equilibrium amount of solute that can dissolve per amount of solvent is the solubility of that solute in that solvent under the specified conditions. The advantage of expressing solubility in this manner is its simplicity, while the disadvantage is that it can strongly depend on the presence of other species in the solvent (for example, the common ion effect).

Solubility constants are used to describe saturated solutions of ionic compounds of relatively low solubility. The solubility constant is a special case of an equilibrium constant. It describes the balance between dissolved ions from the salt and undissolved salt. The solubility constant is also "applicable" (i.e. useful) to precipitation, the reverse of the dissolving reaction. As with other equilibrium constants, temperature can affect the numerical value of solubility constant. The solubility constant is not as simple as solubility, however the value of this constant is generally independent of the presence of other species in the solvent.

The Flory-Huggins solution theory is a theoretical model describing the solubility of polymers. The Hansen Solubility Parameters and the Hildebrand solubility parameters are empirical methods for the prediction of solubility. It is also possible to predict solubility from other physical constants such as the enthalpy of fusion.

The partition coefficient (Log P) is a measure of differential solubility of a compound in a hydrophobic solvent (octanol) and a hydrophilic solvent (water). The logarithm of these two values enables compounds to be ranked in terms of hydrophilicity (or hydrophobicity).

Applications

Solubility is of fundamental importance in a large number of scientific disciplines and practical applications, ranging from ore processing, to the use of medicines, and the transport of pollutants.

Solubility is often said to be one of the "characteristic properties of a substance," which means that solubility is commonly used to describe the substance, to indicate a substance's polarity, to help to distinguish it from other substances, and as a guide to applications of the substance. For example, indigo is described as "insoluble in water, alcohol, or ether but soluble in chloroform, nitrobenzene, or concentrated sulphuric acid".

Solubility of a substance is useful when separating mixtures. For example, a mixture of salt (sodium chloride) and silica may be separated by dissolving the salt in water, and filtering off the undissolved silica. The synthesis of chemical compounds, by the milligram in a laboratory, or by the ton in industry, both make use of the relative solubilities of the desired product, as well as unreacted starting materials, by-products, and side products to achieve separation.

Another example of this is the synthesis of benzoic acid from phenylmagnesium bromide and dry ice. Benzoic acid is more soluble in an organic solvent such as dichloromethane or diethyl ether, and when shaken with this organic solvent in a separatory funnel, will preferentially dissolve in the organic layer. The other reaction products, including the magnesium bromide, will remain in the aqueous layer, clearly showing that separation based on solubility is achieved. This process, known as liquid-liquid extraction, is an important technique in synthetic chemistry.

Solubility of Ionic Compounds in Water

Some ionic compounds (salts) dissolve in water, which arises because of the attraction between positive and negative charges. For example, the salt's positive ions (e.g. Ag^+) attract the partially-negative oxygens in H_2O. Likewise, the salt's negative ions (e.g. Cl^-) attract the partially-positive hydrogens in H_2O. Note: oxygen is partially-negative because it is more electronegative than hydrogen, and vice-versa.

$$AgCl_{(s)}\ Ag^+_{(aq)} + Cl^-_{(aq)}$$

However, there is a limit to how much salt can be dissolved in a given volume of water. This amount is given by the solubility product, K_{sp}. This value depends on the type of salt (AgCl vs. NaCl, for example), temperature, and the common ion effect.

One can calculate the amount of AgCl that will dissolve in 1 litre of water, some algebra is required.

$K_{sp} = [Ag^+] \times [Cl^-]$ (definition of solubility product)

$K_{sp} = 1.8 \times 10^{-10}$ (from a table of solubility products)

$[Ag^+] = [Cl^-]$, in the absence of other silver or chloride salts,

$[Ag^+]^2 = 1.8 \times 10^{-10}$

$[Ag^+] = 1.34 \times 10^{-5}$

The result: 1 litre of water can dissolve 1.34×10^{-5} moles of $AgCl_{(s)}$ at room temperature. Compared with other types of salts, AgCl is poorly soluble in water. In contrast, table salt (NaCl) has a higher K_{sp} and is, therefore, more soluble.

Soluble	*Insoluble*
Group I and NH_4^+ compounds	Carbonates (except Group I, NH_4^+ and uranyl compounds)
Nitrates	Sulphites (except Group I and NH_4^+ compounds)
Acetates (ethanoates) (except Ag^+ compounds)	Phosphates (except Group I and NH_4^+ Compounds)
Chlorides, bromides and iodides (except Ag^+, Pb^{2+}, Cu^+ and Hg_2^{2+})	hydroxides and oxides (except Group I, NH_4^+, Ba^{2+}, Sr^{2+} and Tl^+)
Sulphates (except Ag^+, Pb^{2+}, Ba^{2+}, Sr^{2+} and Ca^{2+})	sulphides (except Group I, Group II and NH_4^+ compounds)

Solubility of Organic Compounds

The principle outlined above under polarity, that *like dissolves like,* is the usual guide to solubility with organic systems. For example, petroleum jelly will dissolve in gasoline because both petroleum jelly and gasoline are hydrocarbons. It will not, on the other hand, dissolve in alcohol or water, since the polarity of these solvents is too high. Sugar will not dissolve in gasoline, since sugar is too polar in comparison with gasoline. A mixture of gasoline and sugar can therefore be separated by filtration, or extraction with water.

Solubility in Non-aqueous Solvents

Most publicly available solubility values are those for solubility in water. Solubility data for non-aqueous solvents is currently being collected via an open notebook science crowdsourcing project.

Solid Solution

This term is often used in the field of metallurgy to refer to the extent that an alloying element will dissolve into the base metal without forming a separate phase. The solubility line (or curve) is the line (or lines) on a phase diagram which give the limits of solute addition. That is, the lines show the maximum amount of a component that can be added to another component and still be in solid solution. In microelectronic fabrication, solid solubility refers to the maximum concentration of impurities one can place into the substrate.

Incongruent Dissolution

Many substances dissolve congruently, i.e., the composition of the solid and the dissolved solute stoichiometrically match. However, some substances may dissolve incongruently, whereby the composition of the solute in solution does not match that of the solid. This solubilisation is accompanied by alteration of the "primary solid" and possibly formation of a secondary solid phase. However, generally, some primary solid also remains and a complex solubility equilibrium establishes. For example, dissolution of albite may result in formation of gibbsite.

$$NaAlSi_3O_8(s) + H^+ + 7H_2O = Na^+ + Al(OH)_3(s) + 3H_4SiO_4.$$

In this case, the solubility of albite is expected to depend on the solid-to-solvent ratio. This kind of solubility is of great importance in geology, where it results in formation of metamorphic rocks.

4

Efficient Fertilizer Use

Concept and Perception

Fertigation is defined as the application of nutrients using an irrigation system by introducing the nutrients into the water flowing through the system. The first reported application of commercial fertilizer through a sprinkler irrigation system was in 1958.

Today, we routinely inject fertilizer solutions and suspensions into irrigation systems via calibrated injection pumps that insure precision over both space and time.

Fertigation has increased dramatically in the past 15 years, particularly for sprinkler and drip systems. For drip systems, the expansion is mostly in horticultural and high value crops.

In agricultural areas with declining water supplies, drip systems have also increased. With increased irrigation, a corresponding increase in fertigation has taken place.

It will continue to grow since such systems result in less water usage and better uniformity and lend themselves to the technique much more readily than the less water-efficient and non-uniform furrow and flood systems being replaced.

Figure: Overhead Sprinkler Irrigation System.

Effective fertigation requires knowledge of certain plant characteristics such as optimum daily nutrient consumption rate and root distribution in the soil. Nutrient characteristics such as solubility and mobility are important and irrigation water quality factors such as pH, mineral content, salinity and nutrient solubility must be considered.

The need for irrigation is the main factor in fertigation because the irrigation system is primarily installed to provide water. The opportunity to fertigate is an added benefit.

Figure: Growers often Plant Vegetables under Plastic Mulch and Fertigate using Drip/trickle Irrigation.

Soil Chemistry

Cropping and fertilization history as well as basic soil chemical properties also need to be considered prior to fertigation. Soil testing by the method recommended in a specific area is the best way to assess current fertility needs.

Particular attention should be given to soil pH and its adjustment to an appropriate range for the crop to be grown.

Proper pH can have a great effect on the availability of residual nutrients in soil as well as those added via fertigation. Any required soil pH adjustments should be made using conventional incorporation techniques. Fertigation is not utilised to alter soil pH's.

Cation exchange capacity is an important consideration in determining the quantity of cations that will be retained during fertigation. In order to maintain an acceptable concentration of cations in solution, a soil with a low CEC must have its cations replaced via fertigation more often than one with a relatively high CEC, which can hold a greater quantity of ions. Since nutrients are easily added with fertigation, it is most practical on sandy, droughty soils with low CEC. These soils need frequent irrigation and nutrient replenishment.

Since fertigation does not negate classical soil chemistry, fertigated nutrients behave similarly to conventionally applied soluble nutrients followed by rainfall or irrigation. A reminder - only small quantities of nutrients applied by an overhead sprinkler system are absorbed by the foliage.

There is some concern that since nitrate nitrogen (NO_3-N) is highly mobile, it may leach past the root system and reach groundwater. However, leaching of NO3-N in the sands of the Southern Coastal Plain has been evaluated. Georgia researchers compared conventional sidedress for corn with fertigation split 6 to 9 times during the growing season. Soil NO_3-N with depth was evaluated in late-July (near maturity). Soil NO_3-N for the fertigated treatments was greater in the surface and to depths of 2 inches than for the conventional treatment. At a depth of 2 feet, greater concentrations of NO3-N were found for the conventional application. Data such as these support the application of smaller amounts of N more frequently in order to expose lesser quantities to the possibility of leaching.

In contrast to nitrate N, phosphorus (P) fertilizer moves very little in the soil. When broadcast on the soil surface, it will not normally move more than l to 2 inches below the soil surface. Therefore, fertigation of P is highly inefficient when using sprinkler or flood irrigation techniques. However, P fertilizer applied in drip systems may be rather efficient. An advantage of drip irrigation is that plant nutrients can be applied directly in the root zone. Since the majority of roots in drip-irrigated crops are located within the wetting zone, drip applied P will be placed in the soil region containing the highest root density. Therefore, the P applied in this manner is generally used more efficiently by plants than if the same amount of P were surface applied.

Water Quality Considerations

For successful fertigation, careful attention must be given to water quality. The quality of water required depends on several factors, but foremost is the type of fertigation. Drip fertigation requires the highest quality water, i.e., free of suspended solids and microorganisms that plug small orifices in emitters. Only solutions should be used for drip/trickle fertigation. As the orifice size increases, e.g. on sprinkler systems, suspensions can be tolerated.

Precipitation of fertilizers within the irrigation systems can be a very serious problem if solubilities are exceeded. One such problem commonly encountered is with calcium concentrations exceeding about 100 ppm. As calcium concentration increases and phosphates are injected, the probability of precipitation increases. When the water contains very high levels of calcium and magnesium, an injection of ammonium phosphates will result in precipitates deposited on pipe walls as well as in orifices and sprinklers. Often this will completely plug the irrigation system.

If appreciable salts are in the water used for fertigation, one needs to consider the total amount of salts applied to the crop between leaching rains rather than only the amount applied in a single fertigation. Crops vary widely in their tolerance to salts; therefore, the number of irrigations as well as the salt concentration must be considered.

In some areas, high levels of boron (B) in irrigation water is a problem. Some sensitive crops can tolerate only about

1 ppm B. Total irrigation with water containing 1 ppm B applied to sensitive crops on light textured soils should not exceed 3 acre-inches between leaching rains.

Fertilizers for Injection into Irrigation

Three qualities are necessary for a good source of fertilizer for fertigation: (1) it must contain the needed nutrient elements in a form available to plants or in a form that is readily converted (i.e., urea is readily converted to available NH_4- and NO_3-N forms in good soil conditions); (2) it must be uniformly distributed—either uniformly broadcast for sprinkler, furrow, or flood or uniformly distributed among emitters for drip/ trickle; and (3) the chemistry of the application must be such that plants are not burned or stunted and irrigation lines, emitters, or orifices are not restricted or plugged. Considering these limitations, a rather long list of nutrient sources can still be used. In order to provide for the latter two limitations listed, the materials must either be in solution or be in suspension. For drip systems, solution is always needed and only clear liquid should be used.

Nitrogen is the most fertigated element due to high plant nutritional needs, to its great mobility in soil following chemical and biological transformation to NO_3^-N, and because of the many soluble sources of N fertilizers available for fertigation. Many forms of N have been used in fertigation. Urea and UAN solutions are probably most commonly used, but most any soluble form of dry fertilizer can be used. Anhydrous ammonia or any other N fertilizer that has free ammonia should not be applied through sprinkler systems.

Phosphoric acid, and ammonium phosphate solutions can be used as sources of phosphorus in drip systems. Ammonium phosphate solutions are subject to recipitation if injected into water with high calcium and/or magnesium concentrations.

Potassium fertigation is usually accomplished with potassium chloride (muriate of potash) due to its solubility and relative low cost per pound of K. For most crops, foliage burns are not a problem due to the great dilutions in irrigation systems. Relatively pure grades of potassium chloride (white, 62% K_2O equivalent) are normally used in fertigation. Potassium nitrate has been used as a source of K and N, and potassium sulphate as a source of K and S. These K sources are less soluble and cost more per pound of K than potassium chloride.

Sulphur, when needed, can also be provided as ammonium thiosulphate, ammonium sulphate or flowable S. It can be readily mixed with UAN and several other soluble fertilizer grades and injected. Magnesium sulphate (Epson salts) is often used to supply Mg and S.

Micronutrients such as iron, zinc, copper, and manganese may (if applied in forms such as sulfates or chlorides) react with salts in the irrigation water and cause precipitation and clogging. However, many of the micronutrients can be applied as chelates — such as iron or zinc EDTA. Chelates are generally highly water soluble and consequently, will cause little clogging or precipitation.

Using Fertigation Properly and According to Regulations

Fertilizer nutrients have the potential of getting into the irrigation water source if proper operating procedures and safety devices are not installed or maintained by the operator.

Some states require a permit for fertigation. Be sure you are in compliance with your local regulations.

Reputable fertilizer and chemical dealers should be able to provide the type of equipment shown in following figure.

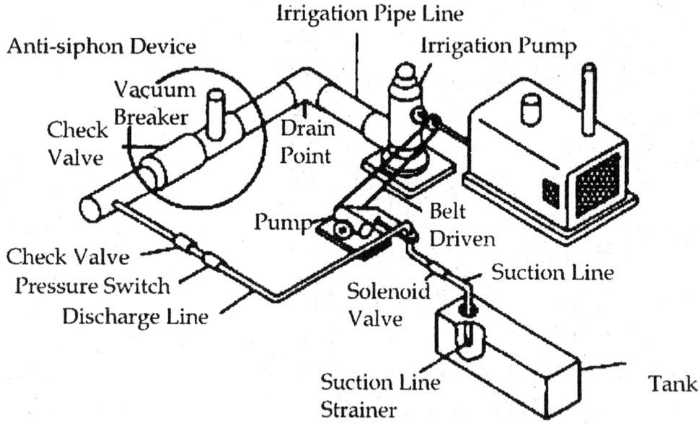

Figure: Safety Devices and Arrangements for Fertigation through an Engine Powered Irrigation System.

Above figure shows a recommended layout for direct injection of nutrients into irrigation pipelines. The following components are part of the system:

1. Backflow prevention device (check valve) in water line upstream of fertilizer injection - this prevents reverse flows from the irrigation system down a well or other water source.
2. Vacuum relief valve - prevents a vacuum from forming upstream of the check valve.
3. Backflow prevention device (automatic, quick-closing check valve) in fertilizer feed line - prevents reverse flows of water or fertilizer into the fertilizer storage tank.
4. Normally closed, solenoid-operated valve - located on the intake side of the injection pump to prevent fertilizer flow during irrigation system shutdown.
5. Electrical interlock for injection systems using electric-driven fertilizer pumps - this ensures that the injection pump will shut down if the irrigation pump does.
6. Low pressure drain valve - to drain water from the pipe between the check valve and the water source, including any leakage past the check valve.

Irrigation Scheduling and Fertigation

Successful moisture management calls for applying the right amount of water at the proper time. The amount of water needed will depend principally on the water-holding capacity of the soil, the soil profile depth, and the crop to be grown.

Research has shown that applying N solutions through sprinkler systems on sandy soils gives greater efficiency than from preplant broadcast and sidedress applications.

This also allows the N to be applied at the time the crop needs it. The data in following tables illustrate the nutrient uptake of N and K for corn and cotton.

Fertilizer applications can be timed with the irrigation to supply nutrients at the most optimum growth stage.

Table: Effect of Multiple Fertilizer Applications on Irrigated Corn.

Fertilizer Schedule	*Time of Last Application**	*No. of Applications*	*Relative Yield %*
Preplant and two sidedress	8	3	72
Bi-weekly starting at emergence	12	7	100

**Weeks after emergence (Source: F. M. Rhoads, University of Florida, Quincy)*

Table: Percent of Nutrients Taken Up by 180-bu. Corn.

Days After Emergence	*%Uptake*	
	N	*K2O*
0-25	8	9
26-50		
(silk)	35	44
51-75	31	31
76-100		
(grain fill)	2	14
101-125	6	2

Table: Percent of Nutrients Taken Up by 1 1/2 Bales Cotton at Various Growth Stages.

Growth Stage	*Nitrogen*	*Potash*
Seedling	6	5
Early Square	14	19
Early Boll	42	36
To Maturity	38	40

Researchers, using sprinkler irrigation, at the Coastal Plain Experiment Station in Georgia achieved increased N efficiency and highest yields when they applied 25% of the N at planting, 22.5% at the six, 12 and 18 leaf stages and 7.5% at tasselling. This technique generally utilises conventional application of N at planting and the remainder via fertigation.

This method improves N uptake early in the season when root systems are small and require high N. Later in the season, this sequence of applications anticipates N requirements while minimising the risks of nitrate leaching.

Fertigation is also sometimes used to apply supplemental applications of potash, sulphur and certain micronutrients. The need to apply such nutrients should be determined by plant analysis.

Frequency and timing of water application have a major impact on yields and operating costs. For the most efficient use of water it is desirable to frequently determine the soil moisture conditions throughout the root zone of the crop being grown. Two proven practical field methods for measuring soil moisture are tensiometers and electrical resistance metres.

Moisture measuring instruments should be installed at three to five locations in the field. The installation site should be representative of the soil types in the field and should be located so that it is convenient to read them on a daily basis.

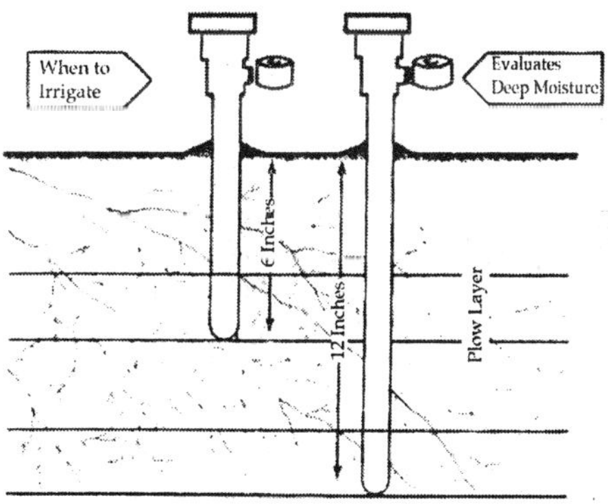

Figure: Installation of Tensiometers in a Corn Field

Since the tensiometers measure the soil water levels, they must be placed in a manner that accurately reflects what conditions the root system is encountering. One should be placed six inches deep and the other 12 inches deep.

The six inch depth tensiometer is used to evaluate when to start irrigating while the twelve inch depth tensiometer is used to evaluate water penetration into the soil profile plus over-or-under watering.

A tensiometer (above figure) reading of less than 20 centibars indicates that the amount of available moisture in the plow layer is favourable. As the readings become higher, the amount of available water decreases, indicating drier conditions and a need to start irrigating.

The response of corn to three selected refill points (readings on the tensiometer) on a loamy fine sand is shown in following table. The highest yields occurred by irrigating corn when the tensiometer reached 20 centibars.

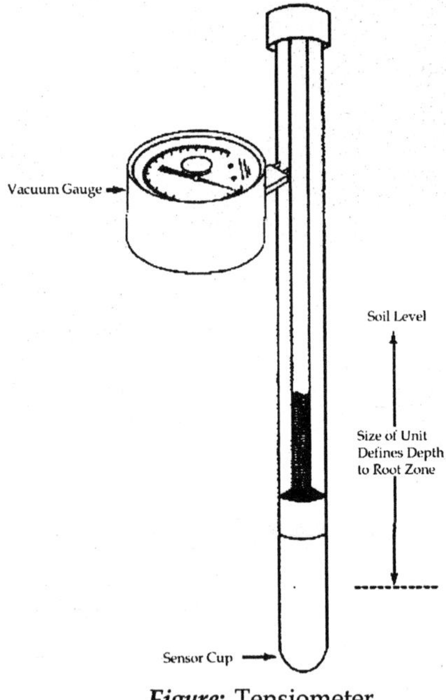

Figure: Tensiometer

Table: Grain Yields of Irrigated Corn at Three Selected Soil/Water Refill Levels

Tensiometer Reading When Irrigated	*Number of Irrigations*	*Total Irrigation Water Applied (IN)*	*YieldBu/A*
20	11	8.7	190
40	6	7.1	175
60	4	4.7	160
–	–	–	115

Source: The University of Georgia Coastal Plain Experiment Station

The type and size of the irrigation system will also influence scheduling. Centre pivot systems that make complete 3600 circles require different schedules than centre pivot systems that do not make complete circles. Cable tow systems require altogether different scheduling. Consult with your irrigation equipment supplier to determine the most efficient schedule for a particular system and field situation.

5

Nitrates and Ammonium

In inorganic chemistry, a nitrate is a salt of nitric acid with an ion composed of one nitrogen and three oxygen atoms (NO^-_3). In organic chemistry the esters of nitric acid and various alcohols are called nitrates.

Chemical Properties

The nitrate ion is a polyatomic ion with the empirical formula NO^-_3 and a molecular mass of 62.0049. It is the conjugate base of nitric acid, consisting of one central nitrogen atom surrounded by three identical oxygen atoms in a trigonal planar arrangement. The nitrate ion carries a formal charge of negative one, where each oxygen carries a $^{-2}D\ _3$ charge while the nitrogen carries a +1 charge, and is commonly used as an example of resonance. Like the isoelectronic carbonate ion, the nitrate ion can be represented by resonance structures:

Almost all inorganic nitrate salts are soluble in water at standard temperature and pressure.

In organic chemistry a nitrate (not to be confused with nitro) is a functional group with general chemical formula $RONO_2$ where R stands for any organic residue. They are the esters of nitric acid and alcohols formed by nitroxylation. Examples are methyl nitrate formed by reaction of methanol and nitric acid, the nitrate of tartaric acid, and the inappropriately named nitroglycerin.

Related Materials

Nitrates should not be confused with nitrites, (NO^-_2) the salts of nitrous acid. Organic compounds containing the nitro functional group (which has the same formula and structure as the nitrate ion save that one of the O^- atoms is replaced by the R group) are known as nitro compounds.

Human Toxicity

Nitrate toxicosis in humans occurs through enterohepatic metabolism of nitrates to ammonia, with nitrite being an intermediate. Nitrites oxidise the iron atoms in haemoglobin from ferrous iron (2+) to ferric iron (3+), rendering it unable to carry oxygen. This condition is called methemoglobinemia and can lead to a lack of oxygen in organ tissue. Methemoglobinemia can be treated with methylene blue, which reduces ferric iron (3+) in affected blood cells back to ferrous iron (2+).

Infants in particular are especially vulnerable to methemoglobinemia due to nitrate metabolising triglycerides present at higher concentrations than at other stages of development. Methemoglobinemia in infants is colloquially know as "blue baby syndrome". Initial exposure is most often caused by high levels of nitrates in drinking water. However, nitrate exposure may also occur if eating for instance vegetables containing high levels of nitrate. Lettuce may contain under growth conditions such as little sunlight, undersupply of the essential micronutrients Molybdenum (Mo) and Iron (Fe) high concentrations of nitrate due to reduced assimilation of nitrate in the plant. High nitrate fertilization also contributes to elevated levels of nitrate in the plant.

Marine Toxicity

In freshwater or estuarine systems close to land, nitrate can reach high levels that can potentially cause the death of fish. While nitrate is much less toxic than ammonia or nitrite, levels over 30 ppm of nitrate can inhibit growth, impair the immune system and cause stress in some aquatic species. However, in light of inherent problems with past protocols on acute nitrate toxicity experiments, the extent of nitrate toxicity has been the subject of recent debate.

In most cases of excess nitrate concentrations in aquatic systems, the primary source is surface run-off from agricultural or landscaped areas which have received excess nitrate fertilizer. These levels of nitrate can also lead to algae blooms, and when nutrients become limiting (such as potassium, phosphate or nitrate) then eutrophication can occur. As well as leading to water anoxia, these blooms may cause other changes to ecosystem function, favouring some groups of organisms over others. Consequently, as nitrates form a component of total dissolved solids, they are widely used as an indicator of water quality.

Nitrates are also a by-product of septic systems. Specifically, they are a naturally occurring chemical that is left after the break down or decomposition of animal or human waste. Water quality may also be affected through ground water resources that have a high number of septic systems in a watershed. Septics leach down into ground water resources or aquifers and supply near by bodies of water. Lakes that rely on ground water are often affected by nitrification through this process.

Ammonium

The *ammonium* cation (also known as ionised ammonia due to its electrical charge) is a positively charged polyatomic cation of the chemical formula NH^{+}_{4}. It has a formula weight of 18.05 and is formed by the protonation of ammonia (NH_3). The resulting ion has a pK_a of 9.25.

Ammonium and aminium are also general names for positively charged or protonated substituted amines and quaternary ammonium cations $N^{+}R_{4}$, where one or more hydrogen atoms arc replaced by organic radical groups (indicated by R).

Chemistry

The ammonium ion is generated by ammonia, a weak base, reacting with Brønsted acids (proton donors):

$$H^+ + :NH_3 \rightarrow NH^+_4$$

The lone electron pair on the nitrogen atom (N) in ammonia is represented as a pair of dots. This electron pair forms the bond with a proton (H^+).

The ammonium ion is a comparatively strong conjugate acid, reacting with Brønsted bases to return to the uncharged ammonia molecule:

$$NH+4 + :B \rightarrow HB + NH_3$$

When ammonia is dissolved in water, a significant amount of it reacts with the hydronium ions in water to give ammonium ions:

$$H_3O^+ + NH_3\ H_2O + NH^+_4$$

The degree to which ammonia forms the ammonium ion depends on the pH of the solution. If the pH is low (there is a high concentration of hydronium ions), the equilibrium shifts to the right: more ammonia molecules are protonated into ammonium ions. If the pH is high (the concentration of hydronium ions is low), the equilibrium shifts to the left: the hydroxide ion abstracts a proton from the ammonium ion, generating ammonia.

Formation of ammonium compounds can also occur in the vapour phase; for example, when ammonia vapour comes in contact with hydrogen chloride vapour, a white cloud of ammonium chloride forms, which eventually settles out as a solid in a thin white layer on surfaces. Ammonium cations resemble alkali metal ions like Na^+ or K^+ and can be found in salts such as ammonium bicarbonate, ammonium chloride, and ammonium nitrate. Most simple ammonium salts are very soluble in water.

Reduction of the ammonium cation gives ammonia gas and hydrogen.

$$2\ NH_{+4} + 2\ e^- \rightarrow 2\ NH_3 + H_2$$

Ammonium ions may dissolve in mercury to form an amalgam. Practically, it may be accomplished by the electrolysis of an ammonium solution with a mercury electrode. This amalgam spontaneously decomposes to give ammonia and hydrogen.

Structure and Bonding

In an ammonium ion, the nitrogen atom forms four covalent bonds (including one coordinate covalent bond), instead of three as in ammonia, forming a structure which is isoelectronic to molecule of methane and so is energetically favourable.

Substituted Ammonium Ions

Each of the hydrogen atoms in the ammonium ion can be substituted with an alkyl group or some other organic group to form a substituted ammonium ion, also called aminium ion. Depending on the number of organic radical groups, it is called a *primary, secondary, tertiary*, or *quaternary* ammonium cation, respectively. Except quaternary ammonium cations, they exist in equilibrium with their respective substituted amines, depending on the pH.

An example of a reaction forming an ammonium ion is that between dimethylamine, $(CH_3)_2NH$, and an acid, to give the dimethylaminium cation, $(CH_3)_2NH_{+2}$:

Quaternary ammonium cations have four organic groups attached to the nitrogen atom. They lack a hydrogen atom bonded to the nitrogen atom which can be abstracted by a base, and so are permanently charged. These cations, such as the tetra-n-butylammonium cation, are sometimes used to replace sodium or potassium ions to increase the overall compound's solubility in organic solvents, based on HSAB principles. They are also used as phase-transfer catalysts for the same reason.

Biology

Ammonium ions are a toxic waste product of the metabolism in animals. In fish and aquatic invertebrates, it is excreted directly into the water. In mammals, sharks, and amphibians, it is converted in the urea cycle to urea, because urea is less toxic and can be stored more efficiently. In birds, reptiles, and terrestrial snails, metabolic ammonium is converted into uric acid, which is solid, and can therefore be excreted with minimal water loss.

Ammonia is toxic to humans in high concentrations, and can cause injury to the mucosal lining of the lung, or alkali burns.

6

Urea

Urea is an organic compound with the chemical formula $(NH_2)_2CO$. This colourless compound is also known by the International Nonproprietary Name (rINN) carbamide, as established by the World Health Organisation. Urea is highly soluble in water and has a pK_a close to zero. Therefore, it is an efficient way for mammals to expel excess nitrogen. The synthesis of this organic compound from an inorganic precursor was important in the development of chemistry.

Discovery

Urea was first discovered in urine in 1773 by the French chemist Hilaire Rouelle.

It was the first organic compound to be artificially synthesised from inorganic starting materials. In 1828 Friedrich Wohler prepared it by treating silver isocyanate with ammonium chloride in a failed attempt to prepare ammonium cyanate:

$$AgNCO + NH_4Cl \rightarrow (NH_2)_2CO + AgCl$$

The results of this experiment implicitly discredited vitalism: the theory that the chemicals of living organisms are fundamentally different from inanimate matter. This insight was important for the development of organic chemistry. His discovery prompted Wohler to write triumphantly to Berzelius: "I must tell you that I can make urea without the use of kidneys, either man or dog. Ammonium cyanate is urea." Wohler's historic preparation of "artificial" urea demonstrated to the scientific world that an organic compound could be synthesised, not only by a living organism, but also in the laboratory. For many, then, Wohler is considered the father of organic chemistry.

It is found in mammalian and amphibian urine as well as in some fish. Birds and reptiles excrete only uric acid, indicative of a different form of nitrogen metabolism that requires less water.

Structure

The urea molecule is planar. Each carbonyl oxygen atom accepts four N-H-O hydrogen bonds. This dense and energetically favourable hydrogen-bond network is probably established at the cost of efficient molecular packing: The structure is quite open, the ribbons forming tunnels with square cross-section. The carbon in urea is described as sp^2 hybridised, the C-N bonds have significant double bond character, and the carbonyl oxygen is basic compared to formaldehyde. Its high solubility is due to extensive hydrogen bonding with water: up to eight hydrogen bonds may form - two from the oxygen atom, one from each hydrogen atom and one from each nitrogen atom.

Physiology

Endogenous Production

Amino acids from ingested food are either used to synthesise proteins and other biomolecules or oxidised to urea and carbon dioxide as a source of energy. The oxidation pathway starts with the removal of the amino group by a transaminase, the amino group is then fed into the urea cycle. Organisms synthesise urea from ammonia because ammonia (a common metabolic waste product) raises pH in cells to toxic levels. Therefore, urea synthesis is necessary even though it costs energy to produce. Urea is neither acidic nor basic, so it is a perfect vehicle for getting rid of nitrogen waste.

Urea production occurs in the liver and is regulated by N-acetylglutamate.

In this cycle, amino groups donated by ammonia and L-aspartate are converted to urea, while L-ornithine, citrulline, L-argininosuccinate, and L-arginine act as intermediates.

Functions

In Humans

Urea is, in essence, a waste product. It is found in and extracted from urine. However, it also plays a very important role in that it helps set up the countercurrent system in the nephrons. The countercurrent system in the nephrons allows for reabsorption of water and critical ions. Urea is reabsorbed in the inner medullary collecting ducts of the nephrons, thus raising the osmolarity in the medullary interstitium surrounding the thin ascending limb of the Loop of Henle. The greater the osmolarity of the medullary interstitium surrounding the thin ascending Loop of Henle, the more water will be reabsorbed out of the renal tubule back into the interstitium (and thus back into the body). Some of the urea from the medullary interstitium that helped set up the countercurrent system will also flow back into the tubule, through urea transporter 2, into the thin ascending limb of the loop of Henle, through the collecting ducts, and eventually out of the body as a component of urine.

It is dissolved in blood (reference range of 2.5 - 7.5 mmol/litre) and excreted by the kidney as a component of urine. In addition, a small amount of urea is excreted (along with sodium chloride and water) in sweat.

Regulation

Control of urea by antidiuretic hormone allows the body to create hyperosmotic urine (urine that has more ions in it—is "more concentrated"—than that same person's blood plasma). Preventing the loss of water in this manner is important if the person's body needs to save water in order to maintain a suitable blood pressure or (more likely,) in order to maintain a suitable concentration of sodium ions in the blood plasma.

Non-humans

Most organisms have to deal with the excretion of nitrogen waste originating from protein and amino acid catabolism. In aquatic organisms the most common form of nitrogen waste is ammonia, while land-dwelling organisms convert the toxic ammonia to either urea or uric acid. In general, birds and saurian reptiles excrete uric acid, whereas the remaining species, including mammals, excrete urea. It is noteworthy that tadpoles excrete ammonia, and shift to urea production during metamorphosis.

Despite the generalisation above, the pathway has been documented not only in mammals and amphibians but in many other organisms as well, including birds, invertebrates, insects, plants, yeast, fungi, and even microorganisms.

Hazards

Urea can be irritating to skin and eyes. Too high concentrations in the blood can cause damage to organs of the body. Low concentrations of urea, such as are found in typical human urine, are not dangerous with additional water ingestion within a reasonable time-frame. Many animals (e.g., dogs) have a much more concentrated urine and it contains a higher urea amount than normal human urine; this can prove dangerous as a source of liquids for consumption in a life-threatening situation (such as in a desert).

It has been found that urea can cause algal blooms to produce toxins, and urea in run-off from fertilizers may play a role in the increase of toxic blooms.

Repeated or prolonged contact with urea in fertilizer form on the skin may cause dermatitis. The substance also irritates the eyes, the skin, and the respiratory tract. The substance decomposes on heating above melting point, producing toxic gases, and reacts violently with strong oxidants, nitrites, inorganic chlorides, chlorites and perchlorates, causing fire and explosion hazard

Synthetic Production

Urea is a nitrogen-containing chemical product that is produced on a scale of some 100,000,000 tons per year worldwide.

For use in industry, urea is produced from synthetic ammonia and carbon dioxide. Urea can be produced as prills, granules, flakes, pellets, crystals, and solutions.

More than 90% of world production is destined for use as a fertilizer. Urea has the highest nitrogen content of all solid nitrogenous fertilizers in common use (46.7%). Therefore, it has the lowest transportation costs per unit of nitrogen nutrient.

Urea is highly soluble in water and is, therefore, also very suitable for use in fertilizer solutions (in combination with ammonium nitrate: UAN), e.g., in 'foliar feed' fertilizers.

Solid urea is marketed as prills or granules. The advantage of prills is that, in general, they can be produced more cheaply than granules, which, because of their narrower particle size distribution, have an advantage over prills if applied mechanically to the soil. Properties such as impact strength, crushing strength, and free-flowing behaviour are, in particular, important in product handling, storage, and bulk transportation.

Commercial Production

Urea is commercially produced from two raw materials, ammonia, and carbon dioxide. Large quantities of carbon dioxide are produced during the manufacture of ammonia from coal or from hydrocarbons such as natural gas and petroleum-derived raw materials. This allows direct synthesis of urea from these raw materials.

The production of urea from ammonia and carbon dioxide takes place in an equilibrium reaction, with incomplete conversion of the reactants. The various urea processes are characterised by the conditions under which urea formation takes place and the way in which unconverted reactants are further processed.

Unconverted reactants can be used for the manufacture of other products, for example ammonium nitrate or sulphate, or they can be recycled for complete conversion to urea in a total-recycle process.

Two principal reactions take place in the formation of urea from ammonia and carbon dioxide. The first reaction of liquid ammonia with dry ice is exothermic:

$$2\ NH_3 + CO_2 \leftrightarrow H_2N\text{-}COONH_4 \text{ (ammonium carbamate)}$$

Whereas the second reaction is endothermic:

$$H_2N\text{-}COONH_4 \leftrightarrow (NH_2)_2CO + H_2O$$

Both reactions combined are exothermic.

The process, developed in 1922, is also called the Bosch-Meiser urea process after its discoverers.

Uses

Agricultural Use: Urea is used as a nitrogen-release fertilizer, as it hydrolyses back to ammonia and carbon dioxide, but its most common impurity, biuret, must be present at less than 2%, as it impairs plant growth. It is also used in many multi-component solid fertilizer formulations. Its action of nitrogen release is due to the conditions favouring the reagent side of the equilibria, which produce urea. If the group of bacteria which "burn" or oxidise urea are not present in the soil then most of the acting ingredient (nitrogen) will not be absorbed by the plants.

Urea is usually spread at rates of between 40 and 300 kg/ha, but actual spreading rates will vary according to farm type and region. It is better to make several small to medium applications at intervals to minimise leaching losses and increase efficient use of the N applied, compared with single heavy applications. During summer, urea should be spread just before, or during rain to reduce possible losses from volatilisation (process wherein nitrogen is lost to the atmosphere as ammonia gas). Urea should not be mixed for any length of time with other fertilizers, as problems of physical quality may result.

Because of the high nitrogen concentration in urea, it is very important to achieve an even spread. The application equipment must be correctly calibrated and properly used. Drilling must not occur on contact with or close to seed, due to the risk of germination damage. Urea dissolves in water for application as a spray or through irrigation systems.

In grain and cotton crops, urea is often applied at the time of the last cultivation before planting. It should be applied into or be incorporated into the soil. In high rainfall areas and on sandy soils (where nitrogen can be lost through leaching) and where good in-season rainfall is expected, urea can be side- or top-dressed during the growing season. Top-dressing is also popular on pasture and forage crops. In cultivating sugarcane, urea is side-dressed after planting, and applied to each ratoon crop.

In irrigated crops, urea can be applied dry to the soil, or dissolved and applied through the irrigation water. Urea will dissolve in its own weight in water, but it becomes increasingly difficult to dissolve as the concentration increases. Dissolving urea in water is endothermic, causing the temperature of the solution to fall when urea dissolves.

As a practical guide, when preparing urea solutions for fertigation (injection into irrigation lines), dissolve no more than 30 kg urea per 100 L water.

In foliar sprays, urea concentrations of 0.5% - 2.0% are often used in horticultural crops. As urea sprays may damage crop foliage, specific advice should be sought before use. Low-biuret grades of urea should be used if urea sprays are to be applied regularly or to sensitive horticultural crops.

Industrial Use

Urea has the ability to form "loose compounds", called clathrates, with many organic compounds. The organic compounds are held in channels formed by interpenetrating helices comprising of hydrogen-bonded urea molecules. This behaviour can be used to separate mixtures, and has been used in the production of aviation fuel and lubricating oils. As the helices are interconnected, all helices in a crystal must have the same "handedness". This is determined when the crystal is nucleated and can thus be forced by seeding. This property has been used to separate racemic mixtures.

Further Commercial Uses:

- A stabiliser in nitrocellulose explosives.
- A reactant in the NO_x-reducing SNCR and SCR reactions in exhaust gases from combustion, for example, from power plants and diesel engines;
- A component of fertilizer and animal feed, providing a relatively cheap source of nitrogen to promote growth.
- A raw material for the manufacture of plastics, to be specific, urea-formaldehyde resin.
- A raw material for the manufacture of various glues (urea-formaldehyde or urea-melamine-formaldehyde); the latter is waterproof and is used for marine plywood.
- An alternative to rock salt in the de-icing of roadways and runways; it does not promote metal corrosion to the extent that salt does.
- An additive ingredient in cigarettes, designed to enhance flavour.
- A browning agent in factory-produced pretzels.
- An ingredient in some hair conditioners, facial cleansers, bath oils, and lotions.
- A reactant in some ready-to-use cold compresses for first-aid use, due to the endothermic reaction it creates when mixed with water.

- A cloud seeding agent, along with salts, to expedite the condensation of water in clouds, producing precipitation.
- An ingredient used in the past to separate paraffins, due to the ability of urea to form clathrates (also called host-guest complexes, inclusion compounds, and adducts).
- A flame-proofing agent (commonly used in dry chemical fire extinguishers as Urea-potassium bicarbonate).
- An ingredient in many tooth whitening products.
- A cream to soften the skin, especially cracked skin on the bottom of one's feet.
- An ingredient in dish soap.
- Along with Ammonium Phosphate, as a Yeast Nutrient, for fermentation of sugars into ethanol.
- To make potassium cyanate.
- A melt agent used in resurfacing snowboarding halfpipes and terrain park features.
- An nutrient used by plankton in ocean nourishment experiments for geoengineering purposes.
- Small amounts added to hide glue lower the temperature at which the glue stays liquid and extends open time.

Laboratory Use

Urea is a powerful protein denaturant. This property can be exploited to increase the solubility of some proteins. For this application, it is used in concentrations up to 10 M. Urea is used to effectively disrupt the noncovalent bonds in proteins. Urea is an ingredient in the synthesis of urea nitrate. Urea nitrate is also a high explosive very similar to ammonium nitrate, however it may even be more powerful because of its complexity. Its VOD is 11,000 ft/s (3,400 m/s) to 15,420 ft/s (4,700 m/s).

Medical Use

Drug Use: Urea is used in topical dermatological products to promote rehydration of the skin. If covered by an occlusive dressing, 40% urea preparations may also be used for nonsurgical debridement of nails. This drug is also used as an earwax removal aid. Like saline, urea injection is used to perform abortions. It is also the main component of an alternative medicinal treatment referred to as urine therapy.

Clinical Diagnosis: The blood urea nitrogen (BUN) test is a measure of the amount of nitrogen in the blood that comes from urea. It is used as a marker of renal function.

Other Diagnostic Use: Isotopically-labelled urea (carbon-14 – radioactive, or carbon-13 – stable isotope) is used in the urea breath test, which is used to detect the presence of the bacteria *Helicobacter pylori* (*H. pylori*) in the stomach and duodenum of humans. The test detects the characteristic enzyme urease, produced by *H. pylori*, by a reaction that produces ammonia from urea. This increases the pH (reduces acidity) of the stomach environment around the bacteria. Similar bacteria species to *H. pylori* can be identified by the same test in animals such as apes, dogs, and cats (including big cats).

Textile Use

Urea in textile laboratories is frequently used both in dyeing and printing as an important auxiliary which provides solubility to the bath and retains some moisture required for the dyeing or printing process.

Storage of Urea Fertilizer

Like most nitrogen products, urea absorbs moisture from the atmosphere. Therefore, it should be stored either in closed/sealed bags on pallets, or, if stored in bulk, under cover with a tarpaulin. As with most solid fertilizers, it should also be stored in a cool, dry, well-ventilated area.

Ionic Liquid

Choline chloride, in mixture with urea, is used as a deep eutectic solvent, a type of ionic liquid.

Automobile Systems

A number of modern diesel engines use the BlueTec system to reduce engine emissions. This consists of an injector containing a water-based urea solution to reduce NO_x emission. The solution is injected into the exhaust system and releases ammonia. This reacts with the nitrogen oxide emissions and is converted into nitrogen and water within the catalytic converter.

Different Ureas

The term *urea* or *carbamide* is also used for the class of chemical compounds sharing the same functional group RR'N-CO-NRR' based on a carbonyl group flanked by two organic amine residues. They can be accessed in the laboratory by reaction of phosgene with primary or secondary amines, proceeding through an isocyanate intermediate. Non-symmetric ureas can be accessed by reaction of primary or secondary amines with an isocyanate. Example of ureas are the compounds carbamide peroxide, allantoin, and Hydantoin. Ureas are closely related to biurets and related in structure to amides, carbamates, diimides, carbodiimides, and thiocarbamides.

Reactions

Urea reacts with alcohols to form urethanes. Urea reacts with malonic esters to make barbituric acids.

Urea Cycle

The urea cycle (also known as the ornithine cycle) is a cycle of biochemical reactions occurring in many animals that produces urea $(NH_2)_2CO$ from ammonia (NH_3). This cycle was the first metabolic cycle discovered (Hans Krebs and Kurt Henseleit, 1932). In mammals, the urea cycle takes place only in the liver.

Functions

Organisms that cannot easily and quickly remove ammonia usually have to convert it to some other substance, like urea or uric acid, which are much less toxic. Insufficiency of the urea cycle occurs in some genetic disorders (inborn errors of metabolism), and in liver failure. The result of liver failure is accumulation of nitrogenous waste, mainly ammonia, which leads to hepatic encephalopathy.

Reactions

The urea cycle consists of five reactions - two mitochondrial and three cytosolic. The cycle converts two amino groups, one from NH_4^+ and one from Asp, and a carbon atom from HCO_3^-, to relatively nontoxic excretion product, urea, at the cost of four "high-energy" phosphate bonds (3 ATP hydrolysed to 2 ADP and one AMP). Orn is the carrier of these carbon and nitrogen atoms.

Reactions of cycle:

Step	*Reactant*	*Product*	*Catalysed by*	*Location*
1.	2ATP + HCO_3^- + NH_4^+	carbamoyl phosphate + 2ADP + P_i	CPS1	mitochondria
2.	carbamoyl phosphate + ornithine	citrulline + P_i	OTC	mitochondria
3.	citrulline + aspartate + ATP	argininosuccinate + AMP + PP_i	ASS	cytosol
4.	argininosuccinate	Arg + fumarate	ASL	cytosol
5.	Arg + H_2O	ornithine + urea	ARG1	cytosol

Overall energy requirement:

- NH_3 + CO_2 + Aspartate + 3 ATP + 2 $H_2O \rightarrow$ urea + Fumarate + 2 ADP + 4 P_i + AMP

Overall equation of the urea cycle:

- $2NH_3$ + CO_2 + 3 ATP + H_2O $\rightarrow$ urea + 2 ADP + 4 P_i + AMP + 2 H

Note that reactions related to the urea cycle also cause the reduction of 2 NADH, so the urea cycle releases slightly more energy than it consumes. These NADH are produced in two ways:

- One NADH molecule is reduced by the enzyme glutamate dehydrogenase in the conversion of glutamate to ammonium and a-ketoglutarate. Glutamate is the non-toxic carrier of amine groups. This provides the ammonium ion used in the initial synthesis of carbamoyl phosphate.
- The fumarate released in the cytosol is converted to malate by cytosolic fumarase. This malate is then converted to oxaloacetate by cytosolic malate dehydrogenase, generating a reduced NADH in the cytosol. Oxaloacetate is one of the keto acids preferred by transaminases, and so will be recycled to aspartate, maintained the flow of nitrogen into the urea cycle.

The two NADH produced can provide energy for the formation of 5 ATP, a net production of one high energy phosphate bond for the urea cycle. However, if gluconeogenesis is underway in the cytosol, the latter reducing equivalent is used to drive the reversal of the GAPDH step instead of generating ATP.

The fate of oxaloacetate is either to produce aspartate via oxidative deamination or to be converted to phosphoenol pyruvate, which is a substrate to glucose.

An excellent way to memorise the Urea Cycle is to remember the phrase "Ordinarily Careless Crappers Are Also Frivolous About Urination." The first letter of each word corresponds to the order in which reactants are combined to give products or intermediates that break apart as one progresses through the cycle.

Regulation

N-Acetylglutamic Acid: The synthesis of carbamoyl phosphate and the urea cycle are dependent on the presence of NAcGlu, which allosterically activates CPS1. Synthesis of NAcGlu by NAGS, is stimulated by Arg – allosteric stimulator of NAGS, and Glu – a product in the transamination reactions and one of NAGS's substrates, both of which are elevated when free amino acids are elevated. So, Arg is not only a substrate for the urea cycle reactions but also serves as an activator for the urea cycle.

Substrate Concentrations

The remaining enzymes of the cycle are controlled by the concentrations of their substrates. Thus, inherited deficiencies in the cycle enzymes other than ARG1 do not result in significant decrease in urea production (the total lack of any cycle enzyme results in death shortly after birth). Rather, the deficient enzyme's substrate builds up, increasing the rate of the deficient reaction to normal.

The anomalous substrate buildup is not without cost, however. The substrate concentrations become elevated all the way back up the cycle to NH_4^+, resulting in hyperammonemia (elevated $[NH_4^+]_P$).

Although the root cause of NH_4^+ toxicity is not completely understood, a high $[NH_4^+]$ puts an enormous strain on the NH_4^+-clearing system, especially in the brain (symptoms of urea cycle enzyme deficiencies include mental retardation and lethargy). This clearing system involves GLUD1 and GLUL, which decrease the 2OG and Glu pools. The brain is most sensitive to the depletion of these pools. Depletion of 2OG decreases the rate of TCAC, whereas Glu is both a neurotransmitter and a precursor to GABA, another neurotransmitter.

Pathology

Anomalies of the urea cycle cause urea cycle disorders:

- ornithine transcarbamoylase deficiency;
- Carbamoyl phosphate synthetase deficiency (Ornithine translocase deficiency);
- Argininosuccinic aciduria;
- Argininemia;
- Hyperornithinemia, hyperammonemia, homocitrullinuria syndrome (HHH syndrome);

- Lysinuric protein intolerance;
- Citrullinemia;
- N-Acetylglutamate synthase deficiency.

Most of them are associated with hyperammonemia.

Renal Urea Handling

Renal urea handling is the part of renal physiology that deals with the reabsorption and secretion of urea.

Urea allows the body to create hyperosmotic urine (urine that has more ions in it—is "more concentrated"—than that same person's blood plasma). Preventing the loss of water in this manner is important if the person's body needs to save water in order to maintain a suitable blood pressure or (more likely,) in order to maintain a suitable concentration of sodium ions in the blood plasma.

About half of the urea filtered (40%) is normally found in the final urine, since there is more reabsorption than secretion along the nephron.

It is regulated by antidiuretic hormone, which controls the amount reabsorbed in the collecting duct system and secreted into the loop of Henle.

Blood Urea Nitrogen

The Blood Urea Nitrogen (BUN) test is a measure of the amount of nitrogen in the blood in the form of urea, and a measurement of renal function. Urea is a substance secreted by the liver, and removed from the blood by the kidneys.

Physiology

The liver produces urea in the urea cycle as a waste product of the digestion of protein. Normal human adult blood should contain between 7 to 21 mg of urea nitrogen per 100 ml (7-21 mg/dL) of blood. Individual laboratories may have different reference ranges, and this is because the procedure may vary.

Interpretation

The most common cause of an elevated BUN, azotemia, is poor kidney function, although a serum creatinine level is a somewhat more specific measure of renal function.

A greatly elevated BUN (>60 mg/dL) generally indicates a moderate-to-severe degree of renal failure. Impaired renal excretion of urea may be due to temporary conditions such as dehydration or shock, or may be due to either acute or chronic disease of the kidneys themselves.

An elevated BUN in the setting of a relatively normal creatinine may reflect a physiological response to a relative decrease of blood flow to the kidney (as seen in heart failure or dehydration) without indicating any true injury to the kidney. However, an isolated elevation of BUN may also reflect excessive formation of urea without any compromise to the kidneys. When the ratio of BUN to creatinine is greater than 20, the patient is suspected of having pre-renal azotemia. This means that the pathologic process is unlikely to be due to intrinsic kidney damage.

Increased production of urea is seen in cases of moderate or heavy bleeding in the upper gastrointestinal tract (e.g. from ulcers). The nitrogenous compounds from the blood are reabsorbed as they pass through the rest of the GI tract and then broken down to urea by the liver. Enhanced metabolism of proteins will also increase urea production, as may be seen with high protein diets, steroid use, burns, or fevers.

A low BUN usually has little significance, but its causes include liver problems, malnutrition (insufficient dietary protein), or excessive alcohol consumption. Overhydration from intravenous fluids can result in a low BUN. Normal changes in renal bloodflow during pregnancy will also lower BUN.

Urea itself is not toxic. This was demonstrated by Johnson *et al.* by adding large amounts of urea to the dialysate of haemodialysis patients for several months and finding no ill effects.. However, BUN is a marker for other nitrogenous waste. Thus, when renal failure leads to a buildup of urea *and other nitrogenous wastes* (uremia), an individual may suffer neurological disturbances such as altered cognitive function (encephalopathy), impaired taste (dysgeusia) or loss of appetite (anorexia). The individual may also suffer from nausea and vomiting, or bleeding from dysfunctional platelets. Prolonged periods of severe uremia may result in the skin taking on a grey discolouration or even forming frank urea crystals (uremic frost) on the skin.

Because multiple variables can interfere with the interpretation of a BUN value, GFR and creatinine clearance are more accurate markers of kidney function. Age, sex, and weight will alter the "normal" range for each individual, including race. In renal failure or chronic kidney disease (CKD), BUN will only be elevated outside "normal" when more than 60% of kidney cells are no longer functioning. Hence, more accurate measures of renal function are generally preferred to assess the clearance for purposes of medication dosing.

Units

BUN is reported as mg/dL in the United States. Elsewhere, the concentration of urea is reported as mmol/L. To convert from mg/dL of blood urea nitrogen to mmol/L of urea, divide by 2.8 (each molecule of urea having 2 nitrogens, each of molar mass 14g/mol)

Urea (in mmol/L) = BUN (in mg/dL of nitrogen) / 2.8 .

Convert BUN to urea in mg/dL by using following formula: Urea= BUN 2.14 MW of urea =60 urea nitrogen: 28 – 60/28.

Methodology

The test as originally carried out was by flame photometry; now chemical colorimetric tests are more widely used. Three methods are common: Diacetyl Monoxime, Urograph and Modified Berthelot Enzymatic methods.

7

The Phosphates

A phosphate, an inorganic chemical, is a salt of phosphoric acid. Inorganic phosphates are mined to obtain phosphorus for use in agriculture and industry. In organic chemistry, a phosphate, or organophosphate, is an ester of phosphoric acid. Organic phosphates are important in biochemistry.

Chemical Properties

The phosphate ion is a polyatomic ion with the empirical formula PO_4^{3-} and a molar mass of 94.973 g/mol. It consists of one central phosphorus atom surrounded by four identical oxygen atoms in a tetrahedral arrangement. The phosphate ion carries a negative three formal charge and is the conjugate base of the hydrogen phosphate ion, HPO_4^{2-}, which is the conjugate base of $H_2PO_4^-$, the dihydrogen phosphate ion, which in turn is the conjugate base of H_3PO_4, phosphoric acid. It is a hypervalent molecule (the phosphorus atom has 10 electrons in its valence shell). Phosphate is also an organophosphorus compound with the formula $OP(OR)_3$. A phosphate salt forms when a positively-charged ion attaches to the negatively-charged oxygen atoms of the ion, forming an ionic compound. Many phosphates are not soluble in water at standard temperature and pressure. The sodium, potassium, rubidium, caesium and ammonium phosphates are all water soluble. Most other phosphates are only slightly soluble or are insoluble in water. As a rule, the hydrogenphosphates and the dihydrogenphosphates are slightly more soluble than the corresponding phosphates. The pyrophosphates are mostly water soluble.

In dilute aqueous solution, phosphate exists in four forms. In strongly-basic conditions, the phosphate ion (PO_4^{3-}) predominates, whereas in weakly-basic conditions, the hydrogen phosphate ion (HPO_4^{2-}) is prevalent. In weakly-acid conditions, the dihydrogen phosphate ion ($H_2PO_4^-$) is most common. In strongly-acid conditions, aqueous phosphoric acid (H_3PO_4) is the main form.

Cellular Function

The addition and removal of phosphate from proteins in all cells is a pivotal strategy in the regulation of metabolic processes.

Phosphate is useful in animal cells as a buffering agent. Phosphate salts that are commonly used for preparing buffer solutions at cell pHs include Na_2HPO_4, NaH_2PO_4, and the corresponding potassium salts.

Mining

Rock phosphate or phosphorite mines are primarily found in:

North America:

- United States of America, especially North Carolina, with lesser deposits in Florida, Idaho and Tennessee.

Africa:

- Morocco mainly near Khouribga and Youssoufia and Western Sahara and Togo.
- Senegal.
- Tunisia.

Middle East:

- Israel.
- Jordan.

- Iraq, at the town of Akashat, near to the Jordanian borders.

Oceania:

- Australia.
- Makatea.
- Nauru.
- Ocean Island.

Potassium

Potassium is a chemical element. It has the symbol K, atomic number 19, and atomic mass 39.0983. Potassium was first isolated from potash. Elemental potassium is a soft silvery-white metallic alkali metal that oxidises rapidly in air and is very reactive with water, generating sufficient heat to ignite the evolved hydrogen.

Potassium in nature occurs only as ionic salt. As such, it is found dissolved in seawater, and as part of many minerals. Potassium ion is necessary for the function of all living cells, and is thus present in all plant and animal tissues. It is found in especially high concentrations in plant cells, and in a mixed diet, it is most highly concentrated in fruits.

In many respects, potassium and sodium are chemically similar, although they have very different functions in organisms in general, and in animal cells in particular.

Occurrence

Potassium is never found free, as it reacts violently with water, which is abundant in nature. As various compounds, potassium makes up about 2.4% of the weight of the Earth's crust and is the seventh most abundant element. As it is very electropositive, potassium metal is difficult to obtain from its minerals.

Free Element

Potassium as such was not known in Roman times, and its names are not Classical Latin but rather neo-Latin. The name kalium was taken from the word "alkali", which came from Arabic *al qaliy* = "the calcined ashes". The name potassium was made from the word "potash", which is English, and originally meant an alkali extracted in a *pot* from the *ash* of burnt wood or tree leaves. The structure of potash was not then known, but is now understood to be mostly potassium carbonate. By heating, the carbonate could be freed of carbon dioxide, leaving "caustic potash," so called because it caused chemical burns in contact with human tissue.

Potassium metal was discovered in 1807 in England by Sir Humphry Davy, who derived it from caustic potash (KOH), by the use of electrolysis of the molten salt with the newly discovered voltaic pile. Before the 18th century, no distinction was made between potassium and sodium. Potassium was the first metal that was isolated by electrolysis. Davy extracted sodium by a similar technique, demonstrating the elements to be different.

Production

Pure potassium metal may be isolated by electrolysis of its hydroxide in a process that has changed little since Davy. Thermal methods also are employed in potassium production, using potassium chloride

Potassium salts such as carnallite, langbeinite, polyhalite, and sylvite form extensive deposits in ancient lake and seabeds, making extraction of potassium salts in these environments commercially viable. The principal source of potassium, potash, is mined in Saskatchewan, California, Germany, New Mexico, Utah, and in other places around the world. It is also found abundantly in the Dead Sea. Three thousand feet below the surface of Saskatchewan are large deposits of potash which are important sources of this element and its salts, with several large mines in operation since the 1960s. Saskatchewan pioneered the use of freezing of wet sands (the Blairmore formation) in order to drive mine shafts through them. The main mining company is the Potash Corporation of Saskatchewan. The oceans are another source of potassium, but the quantity present in a given volume of seawater is relatively low compared with sodium.

Isotopes

There are 24 known isotopes of potassium. Three isotopes occur naturally: ^{39}K (93.3%), ^{40}K (0.0117%) and ^{41}K (6.7%). Naturally occurring ^{40}K decays to stable ^{40}Ar (11.2%) by electron capture and by positron emission, and decays to stable ^{40}Ca (88.8%) by beta decay; ^{40}K has a half-life of 1.250×10^9 years. The decay of ^{40}K to ^{40}Ar enables a commonly used method for dating rocks. The conventional K-Ar dating method depends on the assumption that the rocks contained no argon at the time of formation and that all the subsequent radiogenic argon (i.e., ^{40}Ar) was quantitatively retained. Minerals are dated by measurement of the concentration of potassium and the amount of radiogenic ^{40}Ar that has accumulated. The minerals that are best suited for dating include biotite, muscovite, plutonic/ high grade metamorphic hornblende, and volcanic feldspar; whole rock samples from volcanic flows and shallow instrusives can also be dated if they are unaltered.

Outside of dating, potassium isotopes have been used extensively as tracers in studies of weathering. They have also been used for nutrient cycling studies because potassium is a macronutrient required for life.

^{40}K occurs in natural potassium (and thus in some commercial salt substitutes) in sufficient quantity that large bags of those substitutes can be used as a radioactive source for classroom demonstrations. In healthy animals and people, ^{40}K represents the largest source of radioactivity, greater even than ^{14}C. In a human body of 70 kg mass, about 4,400 nuclei of ^{40}K decay per second.

The activity of natural potassium is 31 Bq/g.

Properties

Physical: Potassium is the second least dense metal; only lithium is less dense. It is a soft, low-melting solid that can easily be cut with a knife. Freshly cut potassium is silvery in appearance, but in air it begins to tarnish towards grey immediately.

In a flame test, potassium and its compounds emit a pale violet colour, which may be masked by the strong yellow emission of sodium if it is also present. Cobalt glass can be used to filter out the yellow sodium colour. Potassium concentration in solution is commonly determined by flame photometry, atomic absorption spectrophotometry, inductively coupled plasma, or ion selective electrodes.

Chemical: Potassium must be protected from air for storage to prevent disintegration of the metal from oxide and hydroxide corrosion. Often samples are maintained under a hydrocarbon medium which does not react with alkali metals, such as mineral oil or kerosene.

Like the other alkali metals, potassium reacts violently with water, producing hydrogen. The reaction is notably more violent than that of lithium or sodium with water, and is sufficiently exothermic that the evolved hydrogen gas ignites.

$$2K(s) + 2H_2O(l) \rightarrow H_2(g) + 2KOH(aq)$$

Because potassium reacts quickly with even traces of water, and its reaction products are nonvolatile, it is sometimes used alone, or as NaK (an alloy with sodium which is liquid at room temperature) to dry solvents prior to distillation. In this role, it serves as a potent desiccant.

Potassium hydroxide reacts strongly with carbon dioxide to produce potassium carbonate, and is used to remove traces of CO_2 from air.

Potassium compounds generally have excellent water solubility, due to the high hydration energy of the K^+ ion. The potassium ion is colourless in water.

Methods of separating potassium by precipitation, sometimes used for gravimetric analysis, include the use of sodium tetraphenylborate, hexachloroplatinic acid, and sodium cobaltinitrite.

Potassium Cations in the Body

Biochemical Function: Potassium cations are important in neuron (brain and nerve) function, and in influencing osmotic balance between cells and the interstitial fluid, with their distribution mediated in all animals (but not in all plants) by the so-called Na+/K+-ATPase pump.

Potassium may be detected by taste because it triggers three of the five types of taste sensations, according to concentration. Dilute solutions of potassium ion taste sweet (allowing moderate concentrations in milk and juices), while higher concentrations become increasingly bitter/alkaline, and finally also salty to the taste. The combined bitterness and saltiness of high potassium content solutions makes high-dose potassium supplementation by liquid drinks a palatability challenge.

Membrane Polarisation: Potassium is also important in allowing muscle contraction and the sending of all nerve impulses in animals through action potentials. By nature of their electrostatic and chemical properties, K^+ ions are larger than Na^+ ions, and ion channels and pumps in cell membranes can distinguish between the two types of ions, actively pumping or passively allowing one of the two ions to pass, while blocking the other.

A shortage of potassium in body fluids may cause a potentially fatal condition known as hypokalemia, typically resulting from diarrhoea, increased diuresis and vomiting. Deficiency symptoms include muscle weakness, paralytic ileus, ECG abnormalities, decreased reflex response and in severe cases respiratory paralysis, alkalosis and cardiac arrhythmia.

Filtration and Excretion

Potassium is an essential mineral micronutrient in human nutrition; it is the major cation (positive ion) inside animal cells, and it is thus important in maintaining fluid and electrolyte balance in the body. Sodium makes up most of the cations of blood plasma at a reference range of about 145 milliequivalents per litre (3345 milligrams) and potassium makes up most of the cell fluid cations at about 150 milliequivalents per litre (4800 milligrams). Plasma is filtered through the glomerulus of the kidneys in enormous amounts, about 180 litres per day. Thus, 602,000 milligrams of sodium and 33,000 milligrams of potassium are filtered each day. All but the 1000-10,000 milligrams of sodium and the 1000-4000 milligrams of potassium likely to be in the diet must be reabsorbed. Sodium must be reabsorbed in such a way as to keep the blood volume exactly right and the osmotic pressure correct; potassium must be reabsorbed in such a way as to keep serum concentration as close as possible to 4.8 milliequivalents (about 190 milligrams) per litre.

Sodium pumps in the kidneys must always operate to conserve sodium. Potassium must sometimes be conserved also, but since the amount of potassium in the blood plasma is very small and the pool of potassium in the cells is about thirty times as large, the situation is not so critical for potassium. Since potassium is moved passively in counter flow to sodium in response to an apparent (but not actual) Donnan equilibrium, the urine can never sink below the concentration of potassium in serum except sometimes by actively excreting water at the end of the processing.

Potassium is secreted twice and reabsorbed three times before the urine reaches the collecting tubules. At that point, it usually has about the same potassium concentration as plasma. If potassium were removed from the diet, there would remain a minimum obligatory kidney excretion of about 200 mg per day when the serum declines to 3.0-3.5 milliequivalents per litre in about one week, and can never be cut off completely. Because it cannot be cut off completely, death will result when the whole body potassium declines to the vicinity of one-half full capacity. At the end of the processing, potassium is secreted one more time if the serum levels are too high.

The potassium moves passively through pores in the cell wall. When ions move through pumps there is a gate in the pumps on either side of the cell wall and only one gate can be open at once. As a result, 100 ions are forced through per second. Pores have only one gate, and there only one kind of ion can stream through, at 10 million to 100 million ions per second. The pores require calcium in order to open although it is thought that the calcium works in reverse by blocking at least one of the pores. Carbonyl groups inside the pore on the amino acids mimics the water hydration that takes place in water solution by the nature of the electrostatic charges on four carbonyl groups inside the pore.

Potassium in the Diet and by Supplement

Adequate Intake: A potassium intake sufficient to support life can generally be guaranteed by eating a variety of foods, especially plant foods. Clear cases of potassium deficiency (as defined by symptoms, signs and a below-normal blood level of the element) are rare in healthy individuals eating a balanced diet. Foods with high sources of potassium include orange juice, potatoes, bananas, avocados, tomatoes, broccoli, soya beans, brown rice, garlic and apricots, although it is also common in most fruits, vegetables and meats.

Optimal Intake: Epidemiological studies and studies in animals subject to hypertension indicate that diets high in potassium can reduce the risk of hypertension and possibly stroke (by a mechanism independent of blood pressure), and a potassium deficiency combined with an inadequate thiamine intake has produced heart disease in rats. With these findings, the question of what is the intake of potassium consistent with optimal health, is debated. For example, the 2004 guidelines of the Institute of Medicine specify a DRI of 4,000 mg of potassium (100 mEq), though most Americans consume only half that amount per day, which would make them formally deficient as regards this particular recommendation. Similarly, in the European Union, particularly in Germany and Italy, insufficient potassium intake is somewhat common.

Medical Supplementation and Disease

Supplements of potassium in medicine are most widely used in conjunction with loop diuretics and thiazides, classes of diuretics which rid the body of sodium and water, but have the side effect of also causing potassium loss in urine. A variety of medical and non-medical supplements are available. Potassium salts such as potassium chloride may be dissolved in water, but the salty/bitter taste of high concentrations of potassium ion make palatable high concentration liquid supplements difficult to formulate. Typical medical supplemental doses range from 10 milliequivalents (400 mg, about equal to a cup of milk or 6 oz. of orange juice) to 20 milliequivalents (800 mg) per dose.

Potassium salts are also available in tablets or capsules, which for therapeutic purposes are formulated to allow potassium to leach slowly out of a matrix, since very high concentrations of potassium ion (which might occur next to a solid tablet of potassium chloride) can kill tissue, and cause injury to the gastric or intestinal mucosa. For this reason, non prescription supplement potassium pills are limited by law in the US to only 99 mg of potassium.

Individuals suffering from kidney diseases may suffer adverse health effects from consuming large quantities of dietary potassium. End stage renal failure patients undergoing therapy by renal dialysis must observe strict dietary limits on potassium intake, since the kidneys control potassium excretion, and buildup of blood concentrations of potassium (hyperkalemia) may trigger fatal cardiac arrhythmia.

Applications

About 93% of the world potassium production was consumed by the fertilizer industry.

Biological Applications

Potassium ions are an essential component of plant nutrition and are found in most soil types. Its primary use in agriculture, horticulture and hydroponic culture is as a fertilizer as the chloride (KCl), sulphate (K_2SO_4) or nitrate (KNO_3).

In animal cells, potassium ions are vital to keeping cells alive.

Food Applications

Potassium ion is a nutrient necessary for human life and health. Potassium chloride is used as a substitute for table salt by those seeking to reduce sodium intake so as to control hypertension. The USDA lists tomato paste, orange juice, beet greens, white beans, potatoes, bananas and many other good dietary sources of potassium, ranked according to potassium content per measure shown.

Potassium sodium tartrate, or Rochelle salt ($KNaC_4H_4O_6$) is the main constituent of baking powder. Potassium bromate ($KBrO_3$) is a strong oxidiser, used as a flour improver (E924) to improve dough strength and rise height.

The sulphite compound, potassium bisulphite ($KHSO_3$) is used as a food preservative, for example in wine and beer-making (but not in meats). It is also used to bleach textiles and straw, and in the tanning of leathers.

Non-dietary uses of potassium chloride include its use to stop the heart, e.g. in cardiac surgery and in a solution used in executions by lethal injection.

Industrial Applications

Potassium vapour is used in several types of magnetometers. An alloy of sodium and potassium, NaK (usually pronounced "nack"), that is liquid at room temperature, is used as a heat-transfer medium. It can also be used as a desiccant for producing dry and air-free solvents.

Potassium metal reacts vigorously with all of the halogens to form the corresponding potassium halides, which are white, water-soluble salts with cubic crystal morphology. Potassium bromide (KBr), potassium iodide (KI) and potassium chloride (KCl) are used in photographic emulsion to make the corresponding photosensitive silver halides.

Potassium hydroxide KOH is a strong base, used in industry to neutralise strong and weak acids and thereby finding uses in pH control and in the manufacture of potassium salts. Potassium hydroxide is also used to saponify fats and oils and in hydrolysis reactions, for example of esters and in industrial cleaners.

Potassium nitrate KNO_3 or saltpeter is obtained from natural sources such as guano and evaporites or manufactured by the Haber process and is the oxidant in gunpowder (black powder) and an important agricultural fertilizer. Potassium cyanide KCN is used industrially to dissolve copper and precious metals particularly silver and gold by forming complexes; applications include gold mining, electroplating and electroforming of these metals. It is also used in organic synthesis to make nitriles. Potassium carbonate K_2CO_3, also known as potash, is used in the manufacture of glass and soap and as a mild desiccant.

Potassium chromate (K_2CrO_4) is used in inks, dyes, and stains (bright yellowish-red colour), in explosives and fireworks, in safety matches, in the tanning of leather and in fly paper. Potassium fluorosilicate (K_2SiF_6) is used in specialised glasses, ceramics, and enamels. Potassium sodium tartrate, or Rochelle salt ($KNaC_4H_4O_6$) is used in the silvering of mirrors.

The superoxide KO_2 is an orange-coloured solid used as a portable source of oxygen and as a carbon dioxide absorber. It is useful in portable respiration systems. It is widely used in submarines and spacecraft as it takes less volume than O_2 (g).

$$4KO_2 + 2CO_2 \rightarrow 2K_2CO_3 + 3O_2$$

Potassium chlorate $KClO_3$ is a strong oxidant, used in percussion caps and safety matches and in agriculture as a weedkiller. Glass may be treated with molten potassium nitrate KNO_3 to make toughened glass, which is much stronger than regular glass.

Precautions: Potassium reacts very violently with water producing hydrogen gas which then usually catches fire. Potassium is usually kept under a hydrocarbon oil such as mineral oil or kerosene to stop the metal from reacting with water vapour present in the air. Unlike lithium and sodium, however, potassium should not be stored under oil indefinitely.

If stored longer than 6 months to a year, dangerous shock-sensitive peroxides can form on the metal and under the lid of the container, which can detonate upon opening. It is recommended that potassium, rubidium or caesium not be stored for longer than three months unless stored in an inert (oxygen free) atmosphere, or under vacuum.

As potassium reacts with water to produce highly flammable hydrogen gas, a potassium fire is only exacerbated by the addition of water, and only a few dry chemicals are effective for putting out such a fire.

Potassium also produces potassium hydroxide (KOH) in the reaction with water. Potassium hydroxide is a strong alkali and so is a caustic hazard, causing burns.

Due to the highly reactive nature of potassium metal, it must be handled with great care, with full skin and eye protection being used and preferably an explosive resistant barrier between the user and the potassium.

8

The Potassium

Potassium (K) has 24 known isotopes. Three isotopes occur naturally: ^{39}K (93.3%), ^{40}K (0.012%) and ^{41}K (6.7%). The standard atomic mass is 39.0983(1) u.

Naturally occurring ^{40}K decays to stable ^{40}Ar (11.2%) by electron capture and by positron emission, and decays to stable ^{40}Ca (88.8%) by beta decay; ^{40}K has a half-life of 1.250×10^9 years. The decay of ^{40}K to ^{40}Ar enables a commonly used method for dating rocks.

The conventional K-Ar dating method depends on the assumption that the rocks contained no argon at the time of formation and that all the subsequent radiogenic argon (i.e., ^{40}Ar) was quantitatively retained.

Minerals are dated by measurement of the concentration of potassium and the amount of radiogenic ^{40}Ar that has accumulated.

The minerals that are best suited for dating include biotite, muscovite, plutonic/high grade metamorphic hornblende, and volcanic feldspar; whole rock samples from volcanic flows and shallow instrusives can also be dated if they are unaltered.

Table

Nuclide Symbol	Z(p)	N(n)	Isotopic Mass (u)	Half-life	Nuclear Spin	Representative Isotopic Composition (mole fraction)
	Excitation Energy					
^{32}K	19	13	32.02192(54)#		1+#	
^{32m}K	950(100)# keV			?	4+#	
^{33}K	19	14	33.00726(21)#	<25 ns	(3/2+)#	
^{34}K	19	15	33.99841(32)#	<40 ns	1+#	
^{35}K	19	16	34.988010(21)	178(8) ms	3/2+	
^{36}K	19	17	35.981292(8)	342(2) ms	2+	
^{37}K	19	18	36.97337589(10)	1.226(7) s	3/2+	
^{38}K	19	19	37.9690812(5)	7.636(18) min	3+	
^{38m1}K		130.50(28) keV		924.2(3) ms	0+	
^{38m2}K		3458.0(2) keV		21.98(11) μs	(7+),(5+)	
^{39}K	19	20	38.96370668(20)	STABLE	3/2+	0.932581(44)
^{40}K	19	21	39.96399848(21)	1.248(3)E+9 a	4-	0.000117(1)
^{40m}K		1643.639(11) keV		336(12) ns	0+	
^{41}K	19	22	40.96182576(21)	STABLE	3/2+	0.067302(44)
^{42}K	19	23	41.96240281(24)	12.360(12) h	2-	
^{43}K	19	24	42.960716(10)	22.3(1) h	3/2+	
^{44}K	19	25	43.96156(4)	22.13(19) min	2-	
^{45}K	19	26	44.960699(11)	17.3(6) min	3/2+	
^{46}K	19	27	45.961977(17)	105(10) s	2(-)	
^{47}K	19	28	46.961678(9)	17.50(24) s	1/2+	
^{48}K	19	29	47.965514(26)	6.8(2) s	(2-)	
^{49}K	19	30	48.96745(8)	1.26(5) s	(3/2+)	
^{50}K	19	31	49.97278(30)	472(4) ms	(0-,1,2-)	
^{51}K	19	32	50.97638(54)#	365(5) ms	3/2+#	
^{52}K	19	33	51.98261(75)#	105(5) ms	(2-)#	
^{53}K	19	34	52.98712(75)#	30(5) ms	(3/2+)#	
^{54}K	19	35	53.99420(97)#	10(5) ms	2-#	
^{55}K	19	36	54.99971(107)#	3 ms	3/2+#	

Notes:

- Values marked # are not purely derived from experimental data, but at least partly from systematic trends. Spins with weak assignment arguments are enclosed in parentheses.
- Uncertainties are given in concise form in parentheses after the corresponding last digits. Uncertainty values denote one standard deviation, except isotopic composition and standard atomic mass from IUPAC which use expanded uncertainties.

Potassium in Biology

Potassium is an essential mineral macronutrient and is the main intracellular ion for all types of cells. It is important in maintaining fluid and electrolyte balance in the body.

Function in the Body

Potassium is the major cation (positive ion) inside animal cells, while sodium is the major cation outside animal cells. The concentration differences of these charged particles causes a difference in electric potential between the inside and outside of cells, known as the membrane potential. The balance between potassium and sodium is maintained by ion pumps in the cell membrane. The cell membrane potential created by potassium and sodium ions allows the cell generate an action potential—a "spike" of electrical discharge. The ability of cells to produce electrical discharge is critical for body functions such as neurotransmission, muscle contraction, and heart function.

Deficiency:

Hypokalemia: A severe shortage of potassium in body fluids may cause a potentially fatal condition known as hypokalemia. Hypokalemia typically results from loss of potassium through diarrhoea, diuresis, or vomiting. Symptoms are related to alterations in membrane potential and cellular metabolism. Symptoms include muscle weakness and cramps, paralytic ileus, ECG abnormalities, intestinal paralysis, decreased reflex response and (in severe cases) respiratory paralysis, alkalosis and arrhythmia.

In rare cases, habitual consumption of large amounts of black licorice has resulted in hypokalemia. Licorice contains a compound (Glycyrrhizin) that increases urinary excretion of potassium.

Insufficient Intake: Although low dietary intake of potassium does not lead to hypokalemia in healthy individuals, many long-term health risks are related to insufficient dietary potassium.

The 2004 guidelines of the Institute of Medicine specify an RDA of 4700 mg of potassium for adults, based on intake levels that have been found to lower blood pressure, reduce salt sensitivity, and minimise the risk of kidney stones. However, most Americans consume only half that amount per day. Similarly, in the European Union, particularly in Germany and Italy, insufficient potassium intake is widespread. Diseases that may be prevented by adequate potassium intake include stroke, osteoporosis, kidney stones, and hypertension.

Food Sources

Eating a variety of foods that contain potassium is the best way to get an adequate amount. Healthy individuals who eat a balanced diet rarely need supplements. Foods with high sources of potassium include orange juice, potatoes, bananas, avocados, apricots, parsnips and turnips, although many other fruits, vegetables, and meats contain potassium.

Side Effects and Toxicity

Gastrointestinal symptoms are the most common side effects of potassium supplements, including nausea, vomiting, abdominal discomfort, and diarrhoea. Taking potassium with meals or taking a microencapsulated form of potassium may reduce gastrointestinal side effects.

Hyperkalemia is the most serious adverse reaction to potassium. Hyperkalemia occurs when potassium builds up faster than the kidneys can remove it. It is most common in individuals with renal failure. Symptoms of hyperkalemia may include tingling of the hands and feet, muscular weakness, and temporary paralysis. The most serious complication of hyperkalemia is the development of an abnormal heart rhythm (arrhythmia), which can lead to cardiac arrest.

Although hyperkalemia is rare in healthy individuals, oral doses greater than 18 grams taken at one time in individuals not accustomed to high intakes can lead to hyperkalemia. All supplements sold in the US contain no more than 99 mg of potassium; a healthy individual would need to consume more than 180 such pills to experience severe health risks.

Deficiency

A deficiency is a lack of something... Example : there is a deficiency of oxygen in the air.

- In mathematics, a deficient number is a number n for which $\sigma(n) < 2n$.
- In medicine there are a variety of nutrient deficiencies:
  - Avitaminosis is a deficiency of vitamins;
  - Boron deficiency (medicine);
  - Chromium deficiency;
  - Iron deficiency (medicine);
  - Iodine deficiency;
  - Magnesium deficiency (medicine);
  - Micronutrient deficiency.
- In construction, a deficiency is an item, or condition that is considered substandard, or below minimum expectations, such as those mandated by either drawings or specifications or the building code or the fire code, and/or any combination of the foregoing. Deficiencies are routinely discussed and dealt with in construction site meetings.
- In genetics, a genetic deletion is also called a Deficiency.
- In Real estate law, a deficiency in the ability to pay off a debt is also called a deficiency judgement or deficiency judgement.

9

The Pesticides

Pesticides are widely used by farmers to control plant pests and enhance production, but chemical pesticides can also cause water quality problems. Pesticides may appear in surface water due to:

- direct application (e.g. aerial spraying or broadcasting over water bodies);
- run-off during rain storms;
- aerial drift (from adjacent fields).

Some pesticides have also been detected in groundwater.

Farmers may use Integrated Pest Management (IPM) techniques (which can include biological pest control) to maintain control over pests, reduce reliance on chemical pesticides, and protect water quality.

There are few safe ways of disposing of pesticide surpluses other than through containment in well managed landfills or by incineration. In some parts of the world, spraying on land is a permitted method of disposal.

A pesticide is a substance or mixture of substances used to kill a pest. A pesticide is any subsyance or mixture of substance intended for: - preventing, destroying, repelling or mitigating any pest. A pesticide may be a chemical substance, biological agent (such as a virus or bacteria), antimicrobial, disinfectant or device used against any pest. Pests include insects, plant pathogens, weeds, molluscs, birds, mammals, fish, nematodes (roundworms) and microbes that compete with humans for food, destroy property, spread or are a vector for disease or cause a nuisance. Although there are benefits to the use of pesticides, there are also drawbacks, such as potential toxicity to humans and other animals. FAO has defined the term of pesticide as:

> Any substance or mixture of substances intended for preventing, destroying or controlling any pest, including vectors of human or animal disease, unwanted species of plants or animals causing harm during or otherwise interfering with the production, processing, storage, transport or marketing of food, agricultural commodities, wood and wood products or animal feedstuffs, or substances which may be administered to animals for the control of insects, arachnids or other pests in or on their bodies. The term includes substances intended for use as a plant growth regulator, defoliant, desiccant or agent for thining fruit or preventing the premature fall of fruit, and substances applied to crops either before or after harvest to protect the commodity from deterioration during storage and transport.

Types of Pesticides

There are multiple ways of classifying pesticides:

- Algicides or algaecides for the control of algae;
- Avicides for the control of birds;
- Bactericides for the control of bacteria;
- Fungicides for the control of fungi and oomycetes;
- Herbicides (e.g. glyphosate) for the control of weeds;
- Insecticides (e.g. organochlorines, organophosphates, carbamates, and pyrethroids) for the control of insects —

these can be ovicides (substances that kill eggs), larvicides (substances that kill larvae) or adulticides (substances that kill adults);

- Miticides or acaricides for the control of mites;
- Molluscicides for the control of slugs and snails;
- Nematicides for the control of nematodes;
- Rodenticides for the control of rodents;
- Virucides for the control of viruses (e.g. H5N1).

Pesticides can also be classed as synthetic pesticides or biological pesticides (biopesticides), although the distinction can sometimes blur.

Broad-spectrum pesticides are those that kill an array of species, while narrow-spectrum, or selective pesticides only kill a small group of species.

A systemic pesticide moves inside a plant following absorption by the plant. With insecticides and most fungicides, this movement is usually upward (through the xylem) and outward. Increased efficiency may be a result. Systemic insecticides which poison pollen and nectar in the flowers may kill needed pollinators such as bees.

Most pesticides work by poisoning pests.

In 2009, the development of a new class of fungicides called paldoxins was announced. These work by taking advantage of natural defence chemicals released by plants called phytoalexins, which fungi then detoxify using enzymes. The paldoxins inhibit the fungi's detoxification enzymes. They are believed to be safer and greener.

Uses

Pesticides are used to control organisms which are considered harmful. For example, they are used to kill mosquitoes that can transmit potentially deadly diseases like west nile virus, yellow fever, and malaria. They can also kill bees, wasps or ants that can cause allergic reactions. Insecticides can protect animals from illnesses that can be caused by parasites such as fleas. Pesticides can prevent sickness in humans that could be caused by mouldy food or diseased produce. Herbicides can be used to clear roadside weeds, trees and brush.

They can also kill invasive weeds in parks and wilderness areas which may cause environmental damage. Herbicides are commonly applied in ponds and lakes to control algae and plants such as water grasses that can interfere with activities like swimming and fishing and cause the water to look or smell unpleasant.

Uncontrolled pests such as termites and mould can damage structures such as houses. Pesticides are used in grocery stores and food storage facilities to manage rodents and insects that infest food such as grain. Each use of a pesticide carries some associated risk. Proper pesticide use decreases these associated risks to a level deemed acceptable by pesticide regulatory agencies such as the United States Environmental Protection Agency (EPA) and the Pest Management Regulatory Agency (PMRA) of Canada.

Pesticides can save farmers' money by preventing crop losses to insects and other pests; in the US, farmers get an estimated fourfold return on money they spend on pesticides. One study found that not using pesticides reduced crop yields by about 10%. Another study, conducted in 1999, found that a ban on pesticides in the United States may result in a rise of food prices, loss of jobs, and an increase in world hunger.

DDT, sprayed on the walls of houses, is an organochloride that has been used to fight malaria since the 1950s. Recent policy statements by the World Health Organisation have given stronger support to this approach. Dr. Arata Kochi, WHO's malaria chief, said, "One of the best tools we have against malaria is indoor residual house spraying. Of the dozen insecticides WHO has approved as safe for house spraying, the most effective is DDT." However, since then, an October 2007 study has linked breast cancer from exposure to DDT prior to puberty. Poisoning may also occur due to use of DDT and other chlorinated hydrocarbons by entering the human food chain when animal tissues are affected. Symptoms include nervous excitement, tremors, convulsions or death. Scientists estimate that DDT and other chemicals in the organophosphate class of pesticides have saved 7 million human lives since 1945 by preventing the transmission of diseases such as malaria, bubonic plague, sleeping sickness, and typhus.

However, DDT use is not always effective, as resistance to DDT was identified in Africa as early as 1955, and by 1972 nineteen species of mosquito worldwide were resistant to DDT. A study for the World Health Organisation in 2000 from Vietnam established that non-DDT malaria controls were significantly more effective than DDT use. The ecological effect of DDT on organisms is an example of bioaccumulation.

History

Since before 2500 BCE, humans have utilised pesticides to protect their crops. The first known pesticide was elemental sulphur dusting used in Sumer about 4,500 years ago. By the 15th century, toxic chemicals such as arsenic, mercury and lead were being applied to crops to kill pests. In the 17th century, nicotine sulphate was extracted from tobacco leaves for use as an insecticide. The 19th century saw the introduction of two more natural pesticides, pyrethrum which is derived from chrysanthemums, and rotenone which is derived from the roots of tropical vegetables.

In 1939, Paul Muller discovered that DDT was a very effective insecticide. It quickly became the most widely used pesticide in the world.

In the 1940s manufacturers began to produce large amounts of synthetic pesticides and their use became widespread. Some sources consider the 1940s and 1950s to have been the start of the "pesticide era." Pesticide use has increased 50-fold since 1950 and 2.3 million tonnes (2.5 million short tons) of industrial pesticides are now used each year. Seventy-five percent of all pesticides in the world are used in developed countries, but use in developing countries is increasing.

In the 1960s, it was discovered that DDT was preventing many fish-eating birds from reproducing, which was a serious threat to biodiversity. Rachel Carson wrote the best-selling book *Silent Spring* about biological magnification. The agricultural use of DDT is now banned under the Stockholm Convention on Persistent Organic Pollutants, but it is still used in some developing nations to prevent malaria and other tropical diseases by spraying on interior walls to kill or repel mosquitoes.

Regulation

In most countries, in order to sell or use a pesticide, it must be approved by a government agency. For example, in the United States, the Environmental Protection Agency (EPA) does so. Complex and costly studies must be conducted to indicate whether the material is safe to use and effective against the intended pest. During the registration process, a label is created which contains directions for the proper use of the material. Based on acute toxicity, pesticides are assigned to a Toxicity Class.

Some pesticides are considered too hazardous for sale to the general public and are designated restricted use pesticides. Only certified applicators, who have passed an exam, may purchase or supervise the application of restricted use pesticides. Records of sales and use are required to be maintained and may be audited by government agencies charged with the enforcement of pesticide regulations.

In Europe, recent EU legislation has been approved banning the use of highly toxic pesticides including those which are carcinogenic, mutagenic or toxic to reproduction, those which are endocrine-disrupting, and those which are persistent, bioaccumulative and toxic (PBT) or very persistent and very bioaccumulative (vPvB). Measures were approved to improve the general safety of pesticides across all EU member states.

In Canada, 154 municipalities and the entire provinces of Quebec and Ontario have now placed restrictions on the cosmetic use of synthetic lawn pesticides as a result of health and environmental concerns. The Ontario provincial government promised on September 24, 2007 to also implement a province-wide ban on the cosmetic use of lawn pesticides, for protecting the public. Medical and environmental groups support such a ban. On April 22, 2008, the Provincial Government of Ontario announced that it will pass legislation that will prohibit, province-wide, the cosmetic use and sale of lawn and garden pesticides.

The Ontario province-wide pesticide ban on lawn pesticides will come into force on Earth Day, April 22, 2009. Over 250 products will be banned for sale and more than 95 pesticide ingredients will be banned for cosmetic uses. The Ontario legislation would also echo Massachusetts law requiring pesticide manufacturers to reduce the toxins they use in production. The Province of Prince Edward Island has also announced that it will pass legislation that bans cosmetic pesticides, starting 2010. On April 3, 2008, the Canadian Cancer Society released opinion poll results conducted by Ipsos Reid, which established that a clear majority of residents in the provinces of British Columbia and Saskatchewan want province-wide cosmetic lawn pesticide bans, and that the majority of respondents believe that cosmetic pesticides are a threat to their health.

Though pesticide regulations differ from country to country, pesticides and products on which they were used are traded across international borders. To deal with inconsistencies in regulations among countries, delegates to a conference of the United Nations Food and Agriculture Organisation adopted an International Code of Conduct on the Distribution and Use of Pesticides in 1985 to create voluntary standards of pesticide regulation for different countries. The Code was updated in 1998 and 2002. The FAO claims that the code has raised awareness about pesticide hazards and decreased the number of countries without restrictions on pesticide use.

Two other efforts to improve regulation of international pesticide trade are the United Nations London Guidelines for the Exchange of Information on Chemicals in International Trade and the United Nations Codex Alimentarius Commission. The former seeks to implement procedures for ensuring that prior informed consent exists between countries buying and selling pesticides, while the latter seeks to create uniform standards for maximum levels of pesticide residues among participating countries. Both initiatives operate on a voluntary basis.

Reading and following label directions is required by law in countries such as the US and in limited parts of the rest of the world.

One study found pesticide self-poisoning the method of choice in one third of suicides worldwide, and recommended, among other things, more restrictions on the types of pesticides that are most harmful to humans.

Environmental Effects

Pesticide use raises a number of environmental concerns. Over 98% of sprayed insecticides and 95% of herbicides reach a destination other than their target species, including non-target species, air, water, bottom sediments and food. Pesticide drift occurs when pesticides suspended in the air as particles are carried by wind to other areas, potentially contaminating them. Pesticides are one of the causes of water pollution, and some pesticides are persistent organic pollutants and contribute to soil contamination.

Health Effects

Pesticides can present danger to consumers, bystanders, or workers during manufacture, transport, or during and after use.

The American Medical Association recommends limiting exposure to pesticides and using safer alternatives:

Particular uncertainty exists regarding the long-term effects of low-dose pesticide exposures. Current surveillance systems are inadequate to characterise potential exposure problems related either to pesticide usage or pesticide-related illnesses...Considering these data gaps, it is prudent...to limit pesticide exposures...and to use the least toxic chemical pesticide or non-chemical alternative.

Point Source Pollution

Farms with large livestock and poultry operations, such as factory farms, can be a major source of point source wastewater. In the United States, these facilities are called *concentrated animal feeding operations* or *confined animal feeding operations* and are being subject to increasing government regulation.

Animal Wastes

The constituents of animal wastewater typically contain:

- Strong organic content – much stronger than human sewage.
- High solids concentration.
- High nitrate and phosphorus content.
- Antibiotics.
- Synthetic hormones.
- Often high concentrations of parasites and their eggs.
- Spores of *Cryptosporidum* (a protozoan) resistant to drinking water treatment processes.
- Spores of *Giardia.*
- Human pathogenic bacteria such as *Brucella* and *Salmonella.*

Animal wastes from cattle can be produced as solid or semisolid manure or as a liquid slurry. The production of slurry is especially common in housed dairy cattle.

Treatment

Whilst solid manure heaps outdoors can give rise to polluting wastewaters from run-off, this type of waste is usually relatively easy to treat by containment and/or covering of the heap.

Animal slurries require special handling and are usually treated by containment in lagoons before disposal by spray or trickle application to grassland. Constructed wetlands are sometimes used to facilitate treatment of animal wastes, as are anaerobic lagoons. Excessive application or application to sodden land or insufficient land area can result in direct run-off to watercourses, with the potential for causing severe pollution. Application of slurries to land overlying aquifers can result in direct contamination or, more commonly, elevation of nitrogen levels as nitrite or nitrate.

The disposal of any wastewater containing animal waste upstream of a drinking water intake can pose serious health problems to those drinking the water because of the highly resistant spores present in many animals that are capable of causing disabling in humans. This risk exists even for very low-level seepage via shallow surface drains or from rainfall run-off.

Some animal slurries are treated by mixing with straws and composted at high temperature to produce a bacteriologically sterile and friable manure for soil improvement.

Piggery Wastes

Piggery waste is comparable to other animal wastes except that many piggery wastes contain elevated levels of copper that can be toxic in the natural environment. Ascarid worms and their eggs are also common and can infect humans if wastewater treatment is ineffective.

Treatment

As for general animal waste, although the liquid fraction of the waste is frequently separated off and reused in the piggery to avoid the prohibitively expensive costs of disposing of a copper-rich liquor.

Silage Liquor

Fresh or wilted grass or other green crops can be made into the semi-fermented product called silage which can be stored and used as winter forage for cattle and sheep. The production of silage often involves the use of an acid conditioner such as sulphuric acid or formic acid. The process of silage making frequently produces a yellow-brown strongly smelling liquid which is very rich in simple sugars, alcohol, short-chain organic acids and silage conditioner.

This liquor is one of the most polluting organic substances known. The volume of silage liquor produced is generally in proportion to the moisture content of the ensiled material.

Treatment

Silage liquor is best treated through prevention by wilting crops well before silage making. Any silage liquor that is produced can be used as part of the food for pigs.

The most effective treatment is by containment in a slurry lagoon and by subsequent spreading on land following substantial dilution with slurry. Containment of silage liquor on its own can cause structural problems in concrete pits because of the acidic nature of silage liquor.

Milking Parlour Wastes

Although milk has a deserved reputation as an important and valuable food product, its presence in wastewaters is highly polluting because of its organic strength, which can lead to very rapid deoxygenation of receiving waters. Milking parlour wastes also contain large volumes of wash-down water, some animal waste together with cleaning and disinfection chemicals.

Treatment

Milking parlour wastes are often treated in admixture with human sewage in a local sewage treatment plant. This ensures that disinfectants and cleaning agents are sufficiently diluted and amenable to treatment.

Running milking wastewaters into a farm slurry lagoon is a possible option although this tends to consume lagoon capacity very quickly. Land spreading is also a treatment option.

Slaughtering Wastes

Wastewater from slaughtering activities is similar to milking parlour waste although considerably stronger in its organic composition and therefore potentially much more polluting.

Treatment

As for milking parlour waste.

Vegetable Washing Water

Washing of vegetables produces large volumes of water contaminated by soil and vegetable pieces. Low levels of pesticides used to treat the vegetables may also be present together with moderate levels of disinfectants such as chlorine.

Treatment

Most vegetable washing waters are extensively recycled with the solids removed by settlement and filtration. The recovered soil can be returned to the land.

Firewater

Although few farms plan for fires, fires are nevertheless more common on farms than on many other industrial premises. Stores of pesticides, herbicides, fuel oil for farm machinery and fertilizers can all help promote fire and can all be present in environmentally lethal quantities in firewater from fire fighting at farms.

Treatment

All farm environmental management plans should allow for containment of substantial quantities of firewater and for its subsequent recovery and disposal by specialist disposal companies.

The concentration and mixture of contaminants in firewater make them unsuited to any treatment method available on the farm. Even land spreading has produced severe taste and odour problems for downstream water supply companies in the past.

Integrated Pest Management

In agriculture, integrated pest management (IPM) is a pest control strategy that uses a variety of complementary strategies including: mechanical devices, physical devices, genetic, biological, cultural management, and chemical management. These methods are done in three stages: prevention, observation, and intervention. It is an ecological approach with a main goal of significantly reducing or eliminating the use of pesticides while at the same time managing pest populations at an acceptable level.

For their leadership in developing and spreading IPM worldwide, Dr. Perry Adkisson and Dr. Ray F. Smith received the 1997 World Food Prize.

Historical Backdrop

Shortly after World War II, when synthetic insecticides became widely available, entomologists in California developed the concept of "supervised insect control." Around the same time, entomologists in cotton-belt states such as Arkansas were advocating a similar approach. Under this scheme, insect control was "supervised" by qualified entomologists, and insecticide applications were based on conclusions reached from periodic monitoring of pest and natural-enemy populations. This was viewed as an alternative to calendar-based insecticide programmes. Supervised control was based on a sound knowledge of the ecology and analysis of projected trends in pest and natural-enemy populations.

Supervised control formed much of the conceptual basis for the "integrated control" that University of California entomologists articulated in the 1950s. Integrated control sought to identify the best mix of chemical and biological controls for a given insect pest. Chemical insecticides were to be used in manner least disruptive to biological control. The term "integrated" was thus synonymous with "compatible." Chemical controls were to be applied only after regular monitoring indicated that a pest population had reached a level (the economic threshold) that required treatment to prevent the population from reaching a level (the economic injury level) at which economic losses would exceed the cost of the artificial control measures.

IPM extended the concept of integrated control to all classes of pests and was expanded to include tactics other than just chemical and biological controls. Artificial controls such as pesticides were to be applied as in integrated control, but these now had to be compatible with control tactics for all classes of pests. Other tactics, such as host-plant resistance and cultural manipulations, became part of the IPM arsenal. IPM added the multidisciplinary element, involving entomologists, plant pathologists, nematologists, and weed scientists.

In the United States, IPM was formulated into national policy in February 1972 when President Nixon directed federal agencies to take steps to advance the concept and application of IPM in all relevant sectors. In 1979, President Carter established an inter-agency IPM Coordinating Committee to ensure development and implementation of IPM practices.

How IPM Works ?

An IPM regime can be quite simple or sophisticated. Historically, the main focus of IPM programmes was on agricultural insect pests. Although originally developed for agricultural pest management, IPM programmes are now developed to encompass diseases, weeds, and other pests that interfere with the management objectives of sites such as residential and commercial structures, lawn and turf areas, and home and community gardens.

An IPM system is designed around five basic components: The US Environmental Protection Agency has a useful set of IPM principles.

1. *Acceptable Pest Levels:* The emphasis is on *eradication*, not *control*. IPM holds that wiping out an entire pest population is often impossible, and the attempt can be more costly, environmentally unsafe, and frequently unachievable. IPM programmes first work to establish acceptable pest levels, called action thresholds, and apply controls if those thresholds are crossed. These thresholds are pest and site specific, meaning that it may be acceptable at one site to have a weed such as white clover, but at another site it may not be acceptable. This stops the pest gaining resistance to chemicals produced by the plant or applied to the crops. If many of the pests are killed then any that have resistance to the chemical will rapidly reproduce forming a resistant population. By not killing all the pests there are some unresistant pests left that will dilute any resistant genes that appear.
2. *Preventive Cultural Practices:* Selecting varieties best for local growing conditions, and maintaining healthy crops, is the first line of defence, together with plant quarantine and 'cultural techniques' such as crop sanitation (*e.g.* removal of diseased plants to prevent spread of infection).
3. *Monitoring:* Regular observation is the cornerstone of IPM. Observation is broken into two steps, first; inspection and second; identification. Visual inspection, insect and spore traps, and other measurement methods and monitoring tools are used to monitor pest levels. Accurate pest identification is critical to a successful IPM programme. Record-keeping is essential, as is a thorough knowledge of the behaviour and reproductive cycles of target pests. Since insects are cold-blooded, their physical development is dependent on the temperature of their environment. Many insects have had their development cycles modelled in terms of

degree days. Monitor the degree days of an environment to determine when is the optimal time for a specific insect's outbreak.

4. *Mechanical Controls:* Should a pest reach an unacceptable level, mechanical methods are the first options to consider. They include simple hand-picking, erecting insect barriers, using traps, vacuuming, and tillage to disrupt breeding.
5. *Biological Controls:* Natural biological processes and materials can provide control, with minimal environmental impact, and often at low cost. The main focus here is on promoting beneficial insects that eat target pests. Biological insecticides, derived from naturally occurring microorganisms (*e.g.*: *Bt*, entomopathogenic fungi and entomopathogenic nematodes), also fit in this category.
6. *Chemical Controls:* Synthetic pesticides are generally only used as required and often only at specific times in a pests life cycle. Many of the newer pesticide groups are derived from plants or naturally occurring substances (*e.g.*: nicotine, pyrethrum and insect juvenile hormone analogues), and further 'biology-based' or 'ecological' techniques are under evaluation.

IPM is applicable to all types of agriculture and sites such as residential and commercial structures, lawn and turf areas, and home and community gardens. Reliance on knowledge, experience, observation, and integration of multiple techniques makes IPM a perfect fit for organic farming (the synthetic chemical option is simply not considered). For large-scale, chemical-based farms, IPM can reduce human and environmental exposure to hazardous chemicals, and potentially lower overall costs of pesticide application material and labour.

1. Proper identification of pest - What is it? Cases of mistaken identity may result in ineffective actions. If

plant damage due to over-watering are mistaken for a fungal infection, a spray may be used needlessly and the plant still dies.

2. Learn pest and host life cycle and biology. At the time you see a pest, it may be too late to do much about it except maybe spray with a pesticide. Often, there is another stage of the life cycle that is susceptible to preventative actions. For example, weeds reproducing from last year's seed can be prevented with mulches. Also, learning what a pest needs to survive allows you to remove these.

3. Monitor or sample environment for pest population - How many are here? Preventative actions must be taken at the correct time if they are to be effective. For this reason, once you have correctly identified the pest, you begin monitoring BEFORE it becomes a problem. For example, in school cafeterias where roaches may be expected to appear, sticky traps are set out before school starts. Traps are checked at regular intervals so you can see them right away and do something before they get out of hand. Some of the things you might want to monitor about pest populations include: Is the pest present/absent? What is the distribution - all over or only in certain spots? Is the pest population increasing or decreasing?

4. Establish action threshold (economic, health or aesthetic) - How many are too many? In some cases, a certain number of pests can be tolerated. Soya beans are quite tolerant of defoliation, so if you have only a few caterpillars in the field and their population is not increasing dramatically, there is no need to do anything. Conversely, there is a point at which you MUST do something. For the farmer, that point is the one at which the cost of damage by the pest is MORE than the cost of control. This is an economic threshold.

Tolerance of pests varies also by whether or not they are a health hazard (low tolerance) or merely a cosmetic damage (high tolerance in a non-commercial situation). Personal tolerances also vary - many people dislike any insect; some people cannot tolerate dandelions in their yards. Different sites may also have varying requirements based on specific areas. White clover may be perfectly acceptable on the sides of a tee box on a golf course, but unacceptable in the fairway where it could cause confusion in the field of play.

5. Choose an appropriate combination of management tactics. For any pest situation, there will be several options to consider. Options include, mechanical or physical control, cultural controls, biological controls and chemical controls. Mechanical or physical controls include picking pests off plants, or using netting or other material to exclude pests such as birds from grapes or rodents from structures. Cultural controls include keeping an area free of conducive conditions by removing or storing waste properly, removing diseased areas of plants properly. Biological controls can be support either through conservation of natural predators or augmentation of natural predators. Augmentative control includes the introduction of naturally occurring predators at either an inundative or inoculative level. An inundative release would be one that seeks to inundate a site with a pest's predator to impact the pest population. An inoculative release would be a smaller number of pest predators to supplement the natural population and provide ongoing control. Chemical controls would include horticultural oils or the application of pesticides such as insecticides and herbicides. A Green Pest Management IPM programme would use pesticides derived from plants, such as botanicals, or other naturally occurring materials.

6. Evaluate results – How did it work? Evaluation is often one of the most important steps. This is the process to review an IPM programme and the results it generated. Asking the following questions is useful: Did your actions have the desired effect? Was the pest prevented or managed to your satisfaction? Was the method itself satisfactory? Were there any unintended side effects? What will you do in the future for this pest situation? Understanding the effectiveness of the IPM programme allows the site manager to make modifications to the IPM plan prior to pests reaching the action threshold and requiring action again.

10

The Insecticides

An insecticide is a pesticide used against insects in all developmental forms. They include ovicides and larvicides used against the eggs and larvae of insects respectively. Insecticides are used in agriculture, medicine, industry and the household. The use of insecticides is believed to be one of the major factors behind the increase in agricultural productivity in the 20th century. Nearly all insecticides have the potential to significantly alter ecosystems; many are toxic to humans; and others are concentrated in the food chain. It is necessary to balance agricultural needs with environmental and health issues when using insecticides.

Classes of Agricultural Insecticides

The classification of insecticides is done in several different ways:

- *Systemic* insecticides are incorporated by treated plants. Insects ingest the insecticide while feeding on the plants.
- *Contact* insecticides are toxic to insects brought into direct contact. Efficacy is often related to the quality of pesticide application, with small droplets (such as aerosols) often improving performance.
- *Natural* insecticides, such as nicotine and pyrethrum, are made by plants as defences against insects. Nicotine

based insecticides have been barred in the US since 2001 to prevent residues from contaminating foods.

- *Inorganic* insecticides are manufactured with metals and include arsenates copper- and fluorine compounds, which are now seldom used, and sulphur, which is commonly used.
- *Organic* insecticides are synthetic chemicals which comprise the largest numbers of pesticides available for use today.
- *Mode of action* - how the pesticide kills or inactivates a pest - is another way of classifying insecticides. Mode of action is important in predicting whether an insecticide will be toxic to unrelated species such as fish, birds and mammals.

Heavy metals, e.g. lead, mercury, arsenic, as well as plant toxins such as nicotine have been used for many years. Heavy metals are poisonous, but they have been used by farmers for quite a while now. Various plants have been used as folk insecticides for centuries, including tobacco and pyrethrum. Some farmers are reporting successfully using spray of crudely fermented alcohol as an effective insecticide.

Organochlorine Compounds

The insectcidal properties of the best known representative of this class of insecticides, DDT, was made by the Swiss Scientist Paul Muller. For this discovery, he was awarded the Nobel Prize for Physiology and Medicine in 1948. DDT was introduced on the market in 1944. With the rise of the modern chemical industry it was possible to make chlorinated hydrocarbons. DDT works by opening the sodium channels in the nerve cells of the insect.

Organophosphates

The next large class developed was the organophosphates, which bind to acetylcholinesterase and other cholinesterases. This results in disruption of nervous impulses, killing the insect or interfering with its ability to carry on normal functions. Organophosphate insecticides and chemical warfare nerve agents (such as sarin, tabun, soman and VX) work in the same way. Organophosphates have an additive toxic effect to wildlife, so multiple exposures to the chemicals amplifies the toxicity.

Carbamate insecticides have similar toxic mechanisms but have a much shorter duration of action and are thus somewhat less toxic.

Pyrethroids

To mimic the insecticidal activity of the natural compound pyrethrum another class of pesticides, pyrethroid pesticides, have been developed. These are nonpersistent and much less acutely toxic than organophosphates and carbamates.

Neonicotinoids

Neonicotinoids are synthetic analogues of the natural insecticide nicotine (with a much lower acute mammalian toxicity and greater field persistence). Broad-spectrum - systemic insecticides with a rapid action (minutes-hours). They are applied as sprays, drenches, seed and soil treatments - often as substitutes for organophosphates and carbamates. Treated insects exhibit leg tremors, rapid wing motion, stylet withdrawal (aphids), disorientated movement, paralysis and death.

Biological Insecticides

Recent efforts to reduce broad spectrum toxins added to the environment have brought biological insecticides back into vogue. An example is the development and increase in use of *Bacillus thuringiensis*, a bacterial disease of Lepidopterans and some other insects. It is used as a larvicide against a wide variety of caterpillars.

Because it has little effect on other organisms, it is considered more environmentally friendly than synthetic pesticides. The toxin from *B. thuringiensis* (Bt toxin) has been incorporated directly into plants through the use of genetic engineering. Other biological insecticides include products based on entomopathogenic fungi (*e.g. Metarhizium anisopliae*), nematodes (*e.g. Steinernema feltiae*) and viruses (*e.g. Cydia pomonella* granulovirus).

Antifeedants

Many plants have evolved substances like polygodial, which prevents insects from eating, but does not kill them directly. The insect often remains nearby, where it dies of starvation. Since antifeedants are nontoxic they would be ideal as insecticides in agriculture. Much agrochemical research is devoted to make them cheap enough for commercial use.

Pollinator Decline

Insecticides can kill bees and may be a cause of pollinator decline, the loss of bees that pollinate plants, Colony Collapse Disorder, in which worker bees from a beehive or Western honey bee colony abruptly disappear. Loss of pollinators will mean a reduction in crop yields. Sublethal doses of insecticides (i.e. imidacloprid and other neonicotinoids) affect foraging behaviour of bees.. However, research into the causes of Colony Collapse Disorder remains inconclusive.

Household Insecticides

Integrated pest management or IPM in the home begins with restricting the availability to insects of three vital commodities: shelter, water and food. If insects become a problem despite such measures, IPM seeks to control them using the safest possible methods, targeting the approach to the particular pest.

Insect repellent, referred to as "bug spray", comes in a plastic bottle or aerosol can. Applied to clothing, arms, legs, and other extremities, the use of these products will tend to ward off nearby insects. This is not an insecticide.

Insecticide used for killing pests—most often insects, and arachnids—primarily comes in an aerosol can, and is sprayed into the air or a nest as a means of killing the animal. Fly sprays will kill house flies, blowflies, ants, cockroaches and other insects and also spiders. Other preparations are granules or liquids that are formulated with bait that is eaten by insects. For many household pests bait traps are available that contain the pesticide and either pheromone or food baits. Crack and crevice sprays are applied into and around openings in houses such as baseboards and plumbing. Pesticides to control termites are often injected into and around the foundations of homes.

Active ingredients of many household insecticides include permethrin and tetramethrin, which act on the nervous system of insects and arachnids.

Bug sprays should be used in well ventilated areas only, as the chemicals contained in the aerosol and most insecticides can be harmful or deadly to humans and pets. All insecticide products including solids, baits and bait traps should be applied such that they are out of reach of wildlife, pets and children.

Individual Insecticides

Chlorinated Hydrocarbons

- Aldrin;
- Chlordane;
- Chlordecone;
- DDT;
- Dieldrin;
- Endosulfan;
- Endrin;
- Heptachlor;
- Hexachlorobenzene;
- Lindane (gamma-Hexachlorocyclohexane);
- Methoxychlor;

- Mirex;
- Pentachlorophenol;
- TDE.

Organophosphates

- Acephate;
- Azinphos-methyl;
- Bensulide;
- Chlorethoxyfos;
- Chlorpyrifos;
- Chlorpyriphos-methyl;
- Diazinon;
- Dichlorvos (DDVP);
- Dicrotophos;
- Dimethoate;
- Disulfoton;
- Ethoprop;
- Fenamiphos;
- Fenitrothion;
- Fenthion;
- Fosthiazate;
- Malathion;
- Methamidophos;
- Methidathion;
- Mevinphos;
- Naled;
- Omethoate;
- Oxydemeton-methyl;
- Parathion;
- Parathion-methyl;

- Phorate;
- Phosalone;
- Phosmet;
- Phostebupirim;
- Phoxim;
- Pirimiphos-methyl;
- Profenofos;
- Terbufos;
- Tetrachlorvinphos;
- Tribufos;
- Trichlorfon.

Carbamates

- Aldicarb;
- Carbofuran;
- Carbaryl;
- Fenoxycarb;
- Methomyl;
- 2-(1-Methylpropyl)phenyl methylcarbamate.

Phenothiazine

Pyrethroids:

- Allethrin;
- Bifenthrin;
- Cypermethrin;
- Deltamethrin;
- Lambda-cyhalothrin;
- Permethrin;
- Resmethrin;
- Tetramethrin;
- Tralomethrin;
- Transfluthrin.

Neonicotinoids:

- Acetamiprid;
- Clothianidin;
- Imidacloprid;
- Nitenpyram;
- Nithiazine;
- Thiacloprid;
- Thiamethoxam.

Plants Derived

- Caffeine;
- Derris (rotenone);
- Anabasine;
- Anethole (mosquito larvae);
- Annonin;
- Asimina (Pawpaw tree seeds) for lice;
- Azadirachtin;
- Carapa;
- Cinnamon leaf oil (very effective for killing mosquito larvae);
- Cinnamaldehyde (very effective for killing mosquito larvae);
- Cinnamyl acetate (kills mosquito larvae);
- Deguelin;
- Derris;
- *Desmodium caudatum* (leaves and roots);
- Eugenol (mosquito larvae);
- Linalool;
- Myristicin;
- Neem (Azadirachtin);

- *Nicotiana rustica* (Nicotine);
- *Peganum harmala,* seeds (smoke from), root;
- Oregano oil kills beetles *Rhizoppertha dominica* (bug found in stored cereal);
- Polyketide;
- Pyrethrum;
- *Quassia* (South American plant genus);
- Tetranortriterpenoid;
- Thymol (controls varroa mites in bee colonies).

Environmental Effects

Effects on Nontarget Species

Some insecticides kill or harm other creatures in addition to those they are intended to kill. For example, birds may be poisoned when they eat food that was recently sprayed with insecticides or when they mistake insecticide granules on the ground for food and eat it.

Sprayed insecticides may drift from the area to which it is applied and into wildlife areas, especially when it is sprayed aerially.

DDT

One of the bigger drivers in the development of new insecticides has been the desire to replace toxic and irksome insecticides. DDT was introduced as a safer alternative to the lead and arsenic compounds. It is the case that when used under the correct conditions that almost any chemical substance is 'safe', but when used under the wrong conditions most insecticides can be a threat to health and/or the environment.

Some insecticides have been banned due to the fact that they are persistent toxins which have adverse effects on animals and/or humans. An oft-quoted case is that of DDT, an example of a widely used (and maybe misused) pesticide, which was brought to public attention by Rachel Carson's book, Silent

Spring. One of the better known impacts of DDT is to reduce the thickness of the egg shells on predatory birds.

The shells sometimes become too thin to be viable, causing reductions in bird populations. This occurs with DDT and a number of related compounds due to the process of bioaccumulation, wherein the chemical, due to its stability and fat solubility, accumulates in organisms' fatty tissues. Also, DDT may biomagnify which causes progressively higher concentrations in the body fat of animals farther up the food chain. The near-worldwide ban on agricultural use of DDT and related chemicals has allowed some of these birds—such as the peregrine falcon—to recover in recent years.

A number of the organochlorine pesticides have been banned from most uses worldwide and globally they are controlled via the Stockholm Convention on persistent organic pollutants. These include: aldrin, chlordane, DDT, dieldrin, endrin, heptachlor, mirex and toxaphene.

11

The Fungicides

Fungicides are chemical compounds or biological organisms used to kill or inhibit fungi or fungal spores. Fungi can cause serious damage in agriculture, resulting in critical losses of yield, quality and profit. Fungicides are used both in agriculture and to fight fungal infections in animals. Chemicals used to control oomycetes, which are not fungi, are also referred to as fungicides as oomycetes use the same mechanisms as fungi to infect plants.

Fungicides can either be contact or systemic. A contact fungicide kills fungi by direct contact; a systemic fungicide has to be absorbed by the affected organism.

Most fungicides that can be bought retail are sold in a liquid form. The most common active ingredient is sulphur, present at 0.08% in weaker concentrates, and as high as 0.5% for more potent fungicides. Fungicides in powdered form are usually around 90% sulphur and are very toxic. Other active ingredients in fungicides include neem oil, rosemary oil, jojoba oil, and the bacterium *Bacillus subtilis*.

Fungicide residues have been found on food for human consumption, mostly from post-harvest treatments. Some fungicides are dangerous to human health, such as vinclozolin, which has now been removed from use.

Natural Fungicides

Plants and other organisms have chemical defences that give them an advantage against microorganisms such as fungi. Some of these compounds can be used as fungicides:

- Tea tree oil.
- Cinnamaldehyde.
- Cinnamon essential oil.
- Jojoba oil.
- Neem oil.
- Rosemary oil
- Milk.

Whole live or dead organisms that are efficient at killing or inhibiting fungi can sometimes be used as fungicides:

- The bacterium *Bacillus subtilis*.
- Kelp (powdered dried kelp is fed to cattle to protect them from fungi in grass).

Resistance

Pathogens respond to the use of fungicides by evolving resistance. In the field several mechanisms of resistance have been identified. The evolution of fungicide resistance can be gradual or sudden. In qualitative or discrete resistance a mutation (normally to a single gene) produces a race of a fungus with a high degree of resistance. Such resistant varieties also tend to show stability, persisting after the fungicide has been removed from the market. For example sugar beet leaf blotch remains resistant to azoles years after they were no longer used for control of the disease. This is because such mutations often have a high selection pressure when the fungicide is used, but there is low selection pressure to remove them in the absence of the fungicide.

In instances where resistance occurs more gradually a shift in sensitivity in the pathogen to the fungicide can be seen. Such resistance is polygenic – an accumulation of many mutation in different genes each having a small additive effect. This type of resistance is known as quantitative or continuous resistance. In this kind of resistance the pathogen population will revert back to a sensitive state if the fungicide is no longer applied.

Little is known about how variations in fungicide treatment affect the selection pressure to evolve resistance to that fungicide. Evidence shows that the doses that provide the most control of the disease also provide the largest selection pressure to acquire resistance, and that lower doses decreased the selection pressure.

In some cases when a pathogen evolves resistance to one fungicide it automatically obtains resistance to others - a phenomenon known as cross resistance. These additional fungicides are normally of the same chemical family or have the same mode of action, or can be detoxified by the same mechanism. Sometimes negative cross resistance occurs, where resistance to one chemical class of fungicides leads to an increase in sensitivity to a different chemical class of fungicides. This has been seen with carbendazim and diethofencarb.

There are also recorded incidences of pathogens evolving multiple drug resistance – resistance to two chemically different fungicides by separate mutation events. For example *Botrytis cinerea* is resistant to both azoles and dicarboximide fungicides.

There are several routes by which pathogens can evolve fungicide resistance. The most common mechanism appears to be alteration of the target site, particular as a defence against single site of action fungicides. For example Black Sigatoka, an economically important pathogen of banana, is resistant to the QoI fungicides, due to a single nucleotide change resulting one amino acid (glycine) being replaced by another (alanine) in the target protein of the QoI fungicides, cytochrome b. This presumably disrupts the binding of the fungicide to the protein, rendering the fungicide ineffective.

Upregulation of target genes can also render the fungicide ineffective. This is seen in DMI resistant strains of *Venturia inaequalis*.

Resistance to fungicides can also be developed by efficient efflux of the fungicide out of the cell. *Septoria tritici* has developed multiple drug resistance using this mechanism. The pathogen had 5 ABC type transporters with overlapping substrate specificities that together work to effectively pump toxic chemicals out of the cell.

In addition to the mechanisms outlined above, fungi may also develop metabolic pathways that circumvent the target protein, or acquire enzymes that enable metabolism of the fungicide to a harmless substance.

Fungicide Resistance Management

The fungicide resistance action committee (FRAC) has several recommended practices to try to avoid the development of fungicide resistance, especially in at-risk fungicides including *Strobilurins* such as azoxystrobin.

Products should not be used in isolation but rather as mixture, or alternate sprays, with another fungicide with a different mechanism of action. The likelihood of the pathogen developing resistance is greatly decreased by the fact that any resistant isolates to one fungicide will hopefully be killed by the other - in other words two mutations would be required rather than just one. The effectiveness of this technique can be demonstrated by Metalaxyl, a phenylamide fungicide. When used as the sole product in Ireland to control potato blight (*Phytophthora infestans*) resistance developed within one growing season. However, in countries like the UK where it was only ever marketed as a mixture resistance problems developed more slowly.

Fungicides should only be applied when absolutely necessary, especially if they are in an at-risk group. Lowering the amount of fungicide in the environment lowers the selection pressure for resistance to develop.

Manufacturers' doses should always be followed. These doses are normally designed to give the right balance between controlling the disease and limiting the risk of resistance development. Higher doses increase the selection pressure for single site mutations that confer resistance, as all strains but those that carry the mutation will be eliminated, and thus the resistant strain will propagate. Lower doses greatly increase the risk of polygenic resistance, as strains that are slightly less sensitive to the fungicide may survive.

It is also recommended that where possible fungicides are only used in a protective manner, rather than to try to cure already infected crops. Far fewer fungicides have curative/ eradicative ability than protectant. Thus, fungicide preparations advertised as having curative action may only have one active chemical; a single fungicide acting in isolation increases the risk of fungicide resistance.

It is better to use an integrative pest management approach to disease control, rather than relying on fungicides alone. This involves the use of resistant varieties and hygienic practises, such as the removal of potato discard piles and stubble on which the pathogen can overwinter, greatly reducing the titre of the pathogen and thus the risk of fungicide resistance development.

12

The Herbicides

A herbicide is substance used to kill unwanted plants. Selective herbicides kill specific targets while leaving the desired crop relatively unharmed. Some of these act by interfering with the growth of the weed and are often synthetic "imitations" of plant hormones. Herbicides used to clear waste ground, industrial sites, railways and railway embankments are non-selective and kill all plant material with which they come into contact. Some plants produce natural herbicides, such as the genus Juglans (walnuts). They are applied in total vegetation control (TVC) programmes for maintenance of highways and railroads. Smaller quantities are used in forestry, pasture systems, and management of areas set aside as wildlife habitat.

Herbicides are widely used in agriculture and in landscape turf management. In the US, they account for about 70% of all agricultural pesticide use.

Historical Backdrop

Prior to the widespread use of chemical herbicides, cultural controls, such as altering soil pH, salinity, or fertility levels, were used to control weeds. Mechanical control (including tillage) was also (and still is) used to control weeds.

The first widely used herbicide was 2, 4-dichloro-phenoxyacetic acid, often abbreviated 2,4-D. It was first commercialised by the Sherwin-Williams Paint company and saw use in the late-1940s. It is easy and inexpensive to manufacture, and kills many broadleaf plants while leaving grasses largely unaffected (although high doses of 2,4-D at crucial growth periods can harm grass crops such as maize or cereals). The low cost of 2,4-D has led to continued usage today and it remains one of the most commonly used herbicides in the world. Like other acid herbicides, current formulations utilise either an amine salt (usually trimethylamine) or one of many esters of the parent compound. These are easier to handle than the acid. 2,4-D exhibits relatively good *selectivity*, meaning, in this case, that it controls a wide number of broadleaf weeds while causing little to no injury to grass crops at normal use rates. A herbicide is termed selective if it affects only certain types of plants, and nonselective if it inhibits a very broad range of plant types. Other herbicides have been more recently developed that achieve higher levels of selectivity than 2,4-D.

The 1950s saw the introduction of the triazine family of herbicides, which includes atrazine, which have current distinction of being the herbicide family of greatest concern regarding groundwater contamination. Atrazine does not break down readily (within a few weeks) after being applied to soils of above neutral pH. Under alkaline soil conditions atrazine may be carried into the soil profile as far as the water table by soil water following rainfall causing the aforementioned contamination. Atrazine is said to have *carryover,* a generally undesirable property for herbicides.

Glyphosate, frequently sold under the brand name Roundup, was introduced in 1974 for non-selective weed control. It is now a major herbicide in selective weed control in growing crop plants due to the development of crop plants that are resistant to it. The pairing of the herbicide with the resistant seed contributed to the consolidation of the seed and chemistry industry in the late-1990s.

Many modern chemical herbicides for agriculture are specifically formulated to decompose within a short period after application. This is desirable as it allows crops which may be affected by the herbicide to be grown on the land in future seasons. However, herbicides with low residual activity (i.e., that decompose quickly) often do not provide season-long weed control.

Classification of Herbicides

Herbicides can be grouped by activity, use, chemical family, mode of action, or type of vegetation controlled.

By activity:

- Contact herbicides destroy only the plant tissue in contact with the chemical. Generally, these are the fastest acting herbicides. They are less effective on perennial plants, which are able to regrow from rhizhomes, roots or tubers.
- Systemic herbicides are translocated through the plant, either from foliar application down to the roots, or from soil application up to the leaves. They are capable of controlling perennial plants and may be slower acting but ultimately more effective than contact herbicides.

By use:

- Soil-applied herbicides are applied to the soil and are taken up by the roots and/or hypocotyl of the target plant. There are three main types of soil-applied herbicides:
  a. Pre-plant incorporated herbicides are soil applied prior to planting and mechanically incorporated into the soil. The objective for incorporation is to prevent dissipation through photo decomposition and/or volatility.
  b. Pre-emergent herbicides are applied to the soil before the crop emerges and prevent germination or early growth of weed seeds.
  c. Post-emergent herbicides are applied after the crop has emerged.

Their classification by mechanism of action (MOA) indicates the first enzyme, protein, or biochemical step affected in the plant following application. The main mechanisms of action are:

- ACCase inhibitors are compounds that kill grasses. Acetyl coenzyme A carboxylase (ACCase) is part of the first step of lipid synthesis. Thus, ACCase inhibitors affect cell membrane production in the meristems of the grass plant. The ACCases of grasses are sensitive to these herbicides, whereas the ACCases of dicot plants are not.
- *ALS Inhibitors:* The acetolactate synthase (ALS) enzyme (also known as acetohydroxyacid synthase, or AHAS) is the first step in the synthesis of the branched-chain amino acids (valine, leucine, and isoleucine). These herbicides slowly starve affected plants of these amino acids which eventually leads to inhibition of DNA synthesis. They affect grasses and dicots alike. The ALS inhibitor family includes sulfonylureas (SUs), imidazolinones (IMIs), triazolopyrimidines (TPs), pyrimidinyl oxybenzoates (POBs), and sulfonylamino carbonyl triazolinones (SCTs). ALS is a biological pathway that exists only in plants and not in animals thus making the ALS-inhibitors among the safest herbicides.
- *EPSPS Inhibitors:* The enolpyruvylshikimate 3-phosphate synthase enzyme EPSPS is used in the synthesis of the amino acids tryptophan, phenylalanine and tyrosine. They affect grasses and dicots alike. Glyphosate (Roundup) is a systemic EPSPS inhibitor but inactivated by soil contact.
- Synthetic auxin inaugurated the era of organic herbicides. They were discovered in the 1940s after a long study of the plant growth regulator auxin. Synthetic auxins mimic this plant hormone. They have several points of action on the cell membrane, and are effective

in the control of dicot plants. 2,4-D is a synthetic auxin herbicide.

- Photosystem II inhibitors reduce electron flow from water to NADPH2+ at the photochemical step in photosynthesis. They bind to the Qb site on the D1 protein, and prevent quinone from binding to this site. Therefore, this group of compounds cause electrons to accumulate on chlorophyll molecules. As a consequence, oxidation reactions in excess of those normally tolerated by the cell occur, and the plant dies. The triazine herbicides (including atrazine) and urea derivatives (diuron) are photosystem II inhibitors.

Organic Herbicides

Almost all herbicides in use today are considered "organic" herbicides in that they contain carbon as a primary molecular component. A notable exception would be the arsenical class of herbicides. Sometimes they are referred to as synthetic organic herbicides. Recently the term "organic" has come to imply products used in organic farming. Under this definition an organic herbicide is one that can be used in a farming enterprise that has been classified as organic. Organic herbicides are expensive and may not be affordable for commercial production. They are much less effective than synthetic herbicides and are generally used along with cultural and mechanical weed control practices.

Organic herbicides include:

- Spices are now effectively used in patented herbicides.
- Vinegar is effective for 5-20% solutions of acetic acid with higher concentrations most effective but mainly destroys surface growth and so respraying to treat regrowth is needed. Resistant plants generally succumb when weakened by respraying.
- Steam has been applied commercially but is now considered uneconomic and inadequate. It kills surface growth but not underground growth and so respraying to treat regrowth of perennials is needed.

- Flame is considered more effective than steam but suffers from the same difficulties.
- D-limonene (citrus oil), which is the active ingredient in Nature's Avenger Organic Herbicide. D-limonene (citrus oil) is a natural degreasing agent that strips the waxy skin or cuticle from weeds, causing dehydration and ultimately death. Environmentally safe, Nature's Avenger Organic Herbicide is registered by the EPA – (US Environmental Protection Agency), and approved for use in organic production by the USDA – (United States Department of Agriculture) and is OMRI – Organic Materials Review Institute.

Application

Most herbicides are applied as water-based sprays using ground equipment. Ground equipment varies in design, but large areas can be sprayed using self-propelled sprayers equipped with a long boom, of 60 to 80 feet (20 to 25 m) with flat fan nozzles spaced about every 20 in (500 mm). Towed, handheld, and even horse-drawn sprayers are also used.

Synthetic organic herbicides can generally be applied aerially using helicopters or airplanes, and can be applied through irrigation systems (chemigation).

Terminology

- Control is the destruction of unwanted weeds, or the damage of them to the point where they are no longer competitive with the crop.
- Suppression is incomplete control still providing some economic benefit, such as reduced competition with the crop.
- Crop Safety, for selective herbicides, is the relative absence of damage or stress to the crop. Most selective herbicides cause some visible stress to crop plants.

Major Herbicides in Use

- 2,4-D, a broadleaf herbicide in the phenoxy group used in turf and in no-till field crop production. Now mainly used in a blend with other herbicides that allow lower rates of herbicides to be used, it is the most widely used herbicide in the world, third most commonly used in the United States. It is an example of synthetic auxin (plant hormone).

- Aminopyralid is a broadleaf herbicide in the pyridine group, used to control broadleaf weeds on grassland, such as docks, thistles and nettles. Notorious for its ability to persist in compost.
- Atrazine, a triazine herbicide used in corn and sorghum for control of broadleaf weeds and grasses. Still used because of its low cost and because it works extrodinarily well on a broad spectrum of weeds common in the US corn belt, Atrazine is commonly used with other herbicides to reduce the over-all rate of atrazine and to lower the for potential groundwater contamination, it is a photosystem II inhibitor.
- Clopyralid is a broadleaf herbicide in the pyridine group, used mainly in turf, rangeland, and for control of noxious thistles. Notorious for its ability to persist in compost. It is another example of synthetic auxin.
- Dicamba, a post-emergent broadleaf herbicide with some soil activity, used on turf and field corn. It is another example of a synthetic auxin.
- Glufosinate ammonium, a broad-spectrum contact herbicide and is used to control weeds after the crop emerges or for total vegetation control on land not used for cultivation.
- Fluroxypyr, a systemic, selective herbicide used for the control of broad-leaved weeds in small grain cereals, maize, pastures, range land and turf. It is a synthetic auxin. In cereal growing, fluroxypyr's key importance is control of cleavers, *Galium aparine*. Other key broad-leaved weeds are also controlled.
- Glyphosate, a systemic non-selective (it kills any type of plant) herbicide used in no-till burndown and for

weed control in crops that are genetically modified to resist its effects. It is an example of an EPSPs inhibitor.

- Imazapyr a non-selective herbicide used for the control of a broad range of weeds including terrestrial annual and perennial grasses and broadleaved herbs, woody species, and riparian and emergent aquatic species.
- Imazapic, a selective herbicide for both the pre- and post-emergent control of some annual and perennial grasses and some broadleaf weeds. Imazapic kills plants by inhibiting the production of branched chain amino acids (valine, leucine, and isoleucine), which are necessary for protein synthesis and cell growth.
- Linuron is a non-selective herbicide used in the control of grasses and broadleaf weeds. It works by inhibiting photosynthesis.
- Metolachlor is a pre-emergent herbicide widely used for control of annual grasses in corn and sorghum; it has displaced some of the atrazine in these uses.
- Paraquat, a non-selective contact herbicide used for no-till burndown and in aerial destruction of marijuana and coca plantings. More acutely toxic to people than any other herbicide in widespread commercial use.
- Pendimethalin, a pre-emergent herbicide widely used to control annual grasses and some broadleaf weeds in a very wide range of crops, including corn, soya beans, wheat, cotton, many tree and vine crops, and many turfgrass species.
- Picloram, a pyridine herbicide mainly used to control unwanted trees in pastures and edges of fields. It is another synthetic auxin.
- Sodium chlorate, a non-selective herbicide, considered phytotoxic to all green plant parts. It can also kill through root absorption.
- Triclopyr, a systemic, foliar herbicide in the pyridine group. It is used to control broadleaf weeds while leaving grasses and conifers unaffected.

Herbicides of Historical Interest

- 2,4,5-Trichlorophenoxyacetic acid (2,4,5-T) was a widely used broadleaf herbicide until being phased out starting in the late-1970s. While 2,4,5-T itself is of only moderate toxicity, the manufacturing process for 2,4,5-T contaminates this chemical with trace amounts of 2,3,7,8-tetrachlorodibenzo-p-dioxin (TCDD). TCDD is extremely toxic to humans. With proper temperature control during production of 2,4,5-T, TCDD levels can be held to about .005 ppm. Before the TCDD risk was well understood, early production facilities lacked proper temperature controls. Individual batches tested later were found to have as much as 60 ppm of TCDD.
- 2,4,5-T was withdrawn from use in the USA in 1983, at a time of heightened public sensitivity about chemical hazards in the environment. Public concern about dioxins was high, and production and use of other (non-herbicide) chemicals potentially containing TCDD contamination was also withdrawn. These included pentachlorophenol (a wood preservative) and PCBs (mainly used as stabilising agents in transformer oil). Some feel that the 2,4,5-T withdrawal was not based on sound science. 2,4,5-T has since largely been replaced by dicamba and triclopyr.
- Agent Orange was a herbicide blend used by the US military in Vietnam between January 1965 and April 1970 as a defoliant. It was a 50/50 mixture of the *n*-butyl esters of 2,4,5-T and 2,4-D. Because of TCDD contamination in the 2,4,5-T component, it has been blamed for serious illnesses in many veterans and Vietnamese people who were exposed to it. However, research on populations exposed to its dioxin contaminant have been inconsistent and inconclusive. Agent Orange often had much higher levels of TCDD than 2,4,5-T used in the US. The name *Agent Orange* is derived from the orange colour-coded stripe used by

the Army on barrels containing the product. It is worth noting that there were other blends of synthetic auxins at the time of the Vietnam War whose containers were recognised by their colours, such as Agent Purple and Agent Pink.

Health Effects

Certain herbicides affect metabolic pathways and systems unique to plants and not found in animals making many modern herbicides among the safest crop protection products having essentially no effect on mammals, birds, amphibians or reptiles.

Some herbicides cause a range of health effects ranging from skin rashes to death. The pathway of attack can arise from intentional or unintentional direct consumption, improper application resulting in the herbicide coming into direct contact with people or wildlife, inhalation of aerial sprays, or food consumption prior to the labelled pre-harvest interval. Under extreme conditions herbicides can also be transported via surface run-off to contaminate distant water sources. Most herbicides decompose rapidly in soils via soil microbial decomposition, hydrolysis, or photolysis. Some herbicides are more persistent with longer soil half-lives. Other alleged health effects include chest pain, headaches, nausea, and fatigue.

All organic and non-organic herbicides must be extensively tested prior to approval for commercial sale and labelling by the Environmental Protection Agency. However, because of the large number of herbicides in use, there is significant concern regarding health effects. Some of the herbicides in use are known to be mutagenic, carcinogenic or teratogenic.

Some herbicides may have therapeutic uses. There is current research into the use of herbicides as an anti-malaria drug that targets the plant-like apicoplast plastid in the malaria-causing parasite *Plasmodium falciparum*.

13

The Antioxidants

Antioxidants are a recent discovery. As cellular metabolism/energy production requires oxygen, potentially damaging (e.g. mutation causing) compounds known as free radicals can form. Most of these are oxidisers (i.e. acceptors of electrons) and some react very strongly. For normal cellular maintenance, growth, and division, these free radicals must be sufficiently neutralised by antioxidant compounds. Some are produced by the human body with adequate precursors (glutathione, Vitamin C) and those that the body cannot produce may only be obtained through the diet through direct sources (Vitamin C in humans, Vitamin A, Vitamin K) or produced by the body from other compounds (Beta-carotene converted to Vitamin A by the body, Vitamin D synthesised from cholesterol by sunlight).

Phytochemicals and their subgroup polyphenols are the majority of antioxidants; about 4,000 are known. Different antioxidants are now known to function in a cooperative network, e.g. vitamin C can reactivate free radical-containing glutathione or vitamin E by accepting the free radical itself, and so on. Some antioxidants are more effective than others at neutralising different free radicals. Some cannot neutralise certain free radicals. Some cannot be present in certain areas of free radical development (Vitamin A is fat-soluble and protects fat areas, Vitamin C is water soluble and protects those areas). When interacting with a free radical, some antioxidants produce a different free radical compound that is less dangerous or more dangerous than the previous compound. Having a variety of antioxidants allows any by-products to be safely dealt with by more efficient antioxidants in neutralising a free radical's butterfly effect.

Phytochemicals

A growing area of interest is the effect upon human health of trace chemicals, collectively called phytochemicals. These nutrients are typically found in edible plants, especially colourful fruits and vegetables, but also other organisms including seafood, algae, and fungi. The effects of phytochemicals increasingly survive rigorous testing by prominent health organisations. One of the principal classes of phytochemicals are polyphenol antioxidants, chemicals which are known to provide certain health benefits to the cardiovascular system and immune system. These chemicals are known to down-regulate the formation of reactive oxygen species, key chemicals in cardiovascular disease.

Perhaps the most rigorously tested phytochemical is zeaxanthin, a yellow-pigmented carotenoid present in many yellow and orange fruits and vegetables. Repeated studies have shown a strong correlation between ingestion of zeaxanthin and the prevention and treatment of age-related macular degeneration (AMD). Less rigorous studies have proposed a correlation between zeaxanthin intake and cataracts. A second carotenoid, lutein, has also been shown to lower the risk of contracting AMD. Both compounds have been observed to collect in the retina when ingested orally, and they serve to protect the rods and cones against the destructive effects of light.

Another carotenoid, beta-cryptoxanthin, appears to protect against chronic joint inflammatory diseases, such as arthritis. While the association between serum blood levels of beta-cryptoxanthin and substantially decreased joint disease has been established, neither a convincing mechanism for such protection nor a cause-and-effect have been rigorously studied. Similarly, a red phytochemical, lycopene, has substantial credible evidence of negative association with development of prostate cancer.

The correlations between the ingestion of some phytochemicals and the prevention of disease are, in some cases, enormous in magnitude.

Even when the evidence is obtained, translating it to practical dietary advice can be difficult and counter-intuitive. Lutein, for example, occurs in many yellow and orange fruits and vegetables and protects the eyes against various diseases. However, it does not protect the eye nearly as well as zeaxanthin, and the presence of lutein in the retina will prevent zeaxanthin uptake. Additionally, evidence has shown that the lutein present in egg yolk is more readily absorbed than the lutein from vegetable sources, possibly because of fat solubility. At the most basic level, the question "should you eat eggs?" is complex to the point of dismay, including misperceptions about the health effects of cholesterol in egg yolk, and its saturated fat content.

As another example, lycopene is prevalent in tomatoes (and actually is the chemical that gives tomatoes their red colour). It is more highly concentrated, however, in processed tomato products such as commercial pasta sauce, or tomato soup, than in fresh "healthy" tomatoes. Yet, such sauces tend to have high amounts of salt, sugar, other substances a person may wish or even need to avoid.

The following table presents phytochemical groups and common sources, arranged by family:

Family	*Sources*	*Possible Benefits*
flavonoids	berries, herbs, vegetables, wine, grapes, tea	general antioxidant, oxidation of LDLs, prevention of arteriosclerosis and heart disease
isoflavones (phytoestrogens)	soy, red clover, kudzu root	general antioxidant, prevention of arteriosclerosis and heart disease, easing symptoms of menopause, cancer prevention

Okinawans, and Adventists, who generally display longevity and "suffer a fraction of the diseases that commonly kill people in other parts of the developed world, and enjoy more healthy years of life." In sum, they offer three sets of 'best practices' to emulate. The rest is up to you. In common with all three groups is to "Eat fruits, vegetables, and whole grains."

The *National Geographic* article noted that an NIH funded study of 34,000 Seventh-day Adventists between 1976 and 1988 "...found that the Adventists' habit of consuming beans, soy milk, tomatoes, and other fruits lowered their risk of developing certain cancers. It also suggested that eating whole grain bread, drinking five glasses of water a day, and, most surprisingly, consuming four servings of nuts a week reduced their risk of heart disease."

The French "Paradox"

It has been discovered that people living in France live longer. Even though they consume more saturated fats than Americans, the rate of heart disease is lower in France than in North America. A number of explanations have been suggested:

- Reduced consumption of processed carbohydrate and other junk foods.
- Regular consumption of red wine.
- More active lifestyles involving plenty of daily exercise, especially walking; the French are much less dependent on cars than Americans are.
- Higher consumption of artificially produced trans-fats by Americans, which has been shown to have greater lipoprotein effects per gram than saturated fat.

However, statistics collected by the World Health Organisation from 1990-2000 show that the incidence of heart disease in France may have been underestimated and in fact be similar to that of neighbouring countries.

Protein

Protein is an important component of every cell in the body. Hair and nails are mostly made of protein. The body uses protein to build and repair tissues. Also protein is used to make enzymes, hormones, and other body chemicals. Protein is an important building block of bones, muscles, cartilage, skin, and blood.

The protein requirement for each individual differs, as do opinions about whether and to what extent physically active people require more protein. The 2005 Recommended Dietary Allowances (RDA), aimed at the general healthy adult population, provide for an intake of 0.8 - 1 grams of protein per kilogram of body weight (according to the BMI formula), with the review panel stating that "no additional dietary protein is suggested for healthy adults undertaking resistance or endurance exercise". Conversely, Di Pasquale (2008), citing recent studies, recommends a minimum protein intake of 2.2 g/kg "for anyone involved in competitive or intense recreational sports who wants to maximise lean body mass but does not wish to gain weight".

Water and Salts

Water is one of the most important nutrients in the sports diet. It helps eliminate food waste products in the body, regulates body temperature during activity and helps with digestion. Maintaining hydration during periods of physical exertion is key to peak performance. While drinking too much water during activities can lead to physical discomfort, dehydration in excess of 2% of body mass (by weight) markedly hinders athletic performance. Additional carbohydrates and protein before, during, and after exercise increase time to exhaustion as well as speed recovery. Dosage is based on work performed, lean body mass, and environmental factors, especially ambient temperature and humidity, maintaining the right amount is key.

14

The Nitrogen

Nitrogen is a chemical element that has the symbol N and atomic number 7 and atomic mass 14.00674μ. Elemental nitrogen is a colourless, odourless, tasteless and mostly inert diatomic gas at standard conditions, constituting 78% by volume of Earth's atmosphere.

Many industrially important compounds, such as ammonia, nitric acid, organic nitrates (propellants and explosives), and cyanides, contain nitrogen. The extremely strong bond in elemental nitrogen dominates nitrogen chemistry, causing difficulty for both organisms and industry in converting the N_2 into useful compounds, and releasing large amounts of energy when these compounds burn or decay back into nitrogen gas.

The element nitrogen was discovered by Daniel Rutherford, a Scottish physician, in 1772. Nitrogen occurs in all living organisms. It is a constituent element of amino acids and thus of proteins, and of nucleic acids (DNA and RNA). It resides in the chemical structure of almost all neurotransmitters, and is a defining component of alkaloids, biological molecules produced by many organisms.

Historical Backdrop

Nitrogen (Latin *nitrogenium*, where *nitrum* (from Greek *nitron*) means "saltpetre", and *genes* means "forming") is formally considered to have been discovered by Daniel Rutherford in 1772, who called it *noxious air* or *fixed air*. That there was a fraction of air that did not support combustion was well known to the late-18th century chemist. Nitrogen was also studied at about the same time by Carl Wilhelm Scheele, Henry Cavendish, and Joseph Priestley, who referred to it as *burnt air* or *phlogisticated air*.

Nitrogen gas was inert enough that Antoine Lavoisier referred to it as "mephetic air" or *azote*, from the Greek word *azotos* meaning "lifeless". Animals died in it, and it was the principal component of air in which animals had suffocated and flames had burned to extinction. Lavoisier's name for nitrogen is used in French and still remains in English in the common names of many compounds, such as hydrazine and compounds of the azide ion.

Compounds of nitrogen were known in the Middle Ages. The alchemists knew nitric acid as *aqua fortis* (strong water). The mixture of nitric and hydrochloric acids was known as *aqua regia* (royal water), celebrated for its ability to dissolve gold (the *King* of metals). The earliest military, industrial and agricultural applications of nitrogen compounds involved uses of saltpeter (sodium nitrate or potassium nitrate), notably in gunpowder, and much later, as fertilizer.

Properties

Nitrogen is a nonmetal, with an electronegativity of 3.04. It has five electrons in its outer shell and is therefore trivalent in most compounds. The triple bond in molecular nitrogen (N_2) is the strongest in nature. The resulting difficulty of converting (N_2) into other compounds, and the ease (and associated high energy release) of converting nitrogen compounds into elemental N_2, have dominated the role of nitrogen in both nature and human economic activities.

At atmospheric pressure molecular nitrogen condenses (liquifies) at 77 K (–195.8 °C) and freezes at 63 K (–210.0 °C) into the beta hexagonal close-packed crystal allotropic form. Below 35.4 K ("237.6 °C) nitrogen assumes the alpha cubic crystal allotropic form. Liquid nitrogen, a fluid resembling water, but with 80.8% of the density (the density of liquid nitrogen at its boiling point is 0.808 g/mL), is a common cryogen.

Unstable allotropes of nitrogen consisting of more than two nitrogen atoms have been produced in the laboratory, like N_3 and N_4. Under extremely high pressures (1.1 million atm) and high temperatures (2000 K), as produced under diamond anvil conditions, nitrogen polymerises into the single bonded diamond crystal structure, an allotrope nicknamed "nitrogen diamond."

Isotopes

There are two stable isotopes of nitrogen: ^{14}N and ^{15}N. By far the most common is ^{14}N (99.634%), which is produced in the CNO cycle in stars. Of the ten isotopes produced synthetically, ^{13}N has a half life of ten minutes and the remaining isotopes have half lives on the order of seconds or less. Biologically-mediated reactions (e.g., assimilation, nitrification, and denitrification) strongly control nitrogen dynamics in the soil. These reactions typically result in ^{15}N enrichment of the substrate and depletion of the product. 0.73% of the molecular nitrogen in Earth's atmosphere is comprised of the isotopologue $^{14}N^{15}N$ and almost all the rest is $^{14}N_2$.

Radioisotope ^{16}N is the dominant radionuclide in the coolant of pressurised water reactors during normal operation. It is produced from ^{16}O (in water) via (n,p) reaction. It has a short half-life of about 7.1 s, but during its decay back to ^{16}O produces high-energy gamma radiation (5 to 7 MeV). Because of this, the access to the primary coolant piping must be restricted during reactor power operation. ^{16}N is one of the main means used to immediately detect even small leaks from the primary coolant to the secondary steam cycle.

Electromagnetic Spectrum

Molecular nitrogen ($^{14}N_2$) is largely transparent to infrared and visible radiation because it is a homonuclear molecule and thus has no dipole moment to couple to electromagnetic radiation at these wavelengths. Significant absorption occurs at extreme ultraviolet wavelengths, beginning around 100 nanometers. This is associated with electronic transitions in the molecule to states in which charge is not distributed evenly between nitrogen atoms. Nitrogen absorption leads to significant absorption of ultraviolet radiation in the Earth's upper atmosphere as well as in the atmospheres of other planetary bodies. For similar reasons, pure molecular nitrogen lasers typically emit light in the ultraviolet range.

Nitrogen also makes a contribution to visible air glow from the Earth's upper atmosphere, through electron impact excitation followed by emission. This visible blue air glow typically results not from molecular nitrogen, but rather from free nitrogen atoms combining with oxygen to form nitric oxide (NO).

Reactions

Nitrogen is generally unreactive at standard temperature and pressure. N_2 reacts spontaneously with few reagents, being resilient to acids and bases as well as oxidants and most reductants. When nitrogen reacts spontaneously with a reagent, the net transformation is often called nitrogen fixation.

Nitrogen reacts with elemental lithium at STP. Lithium burns in an atmosphere of N_2 to give lithium nitride:

$$6\ Li + N_2 \rightarrow 2\ Li_3N$$

Magnesium also burns in nitrogen, forming magnesium nitride.

$$3\ Mg + N_2 \rightarrow Mg_3N_2$$

N_2 forms a variety of adducts with transition metals. The first example of a dinitrogen complex is $[Ru(NH_3)_5(N_2)]^{2+}$. Such compounds are now numerous, other examples include $IrCl(N_2)(PPh_3)_2$, $W(N_2)_2(Ph_2CH_2CH_2PPh_2)_2$, and $[(\eta^5 - C_5Me_4H)_2 Zr]_2 (\mu_2, \eta^2, \eta^2 - N_2)$. These complexes illustrate how N_2 might bind to the metal(s) in nitrogenase and the catalyst for the Haber-

Bosch Process. A catalytic process to reduce N_2 to ammonia with the use of a molybdenum complex in the presence of a proton source was published in 2005.

The starting point for industrial production of nitrogen compounds is the Haber-Bosch process, in which nitrogen is fixed by reacting N_2 and H_2 over a ferric oxide (Fe_3O_4) catalyst at about 500°C and 200 atmospheres pressure. Biological nitrogen fixation in free-living cyanobacteria and in the root nodules of plants also produces ammonia from molecular nitrogen. The reaction, which is the source of the bulk of nitrogen in the biosphere, is catalysed by the nitrogenase enzyme complex which contains Fe and Mo atoms, using energy derived from hydrolysis of adenosine triphosphate (ATP) into adenosine diphosphate and inorganic phosphate (-20.5 kJ/mol).

Occurrence

Nitrogen is the largest single constituent of the Earth's atmosphere (78.082% by volume of dry air, 75.3% by weight in dry air). It is created by fusion processes in stars, and is estimated to be the 7th most abundant chemical element by mass in the universe.

Molecular nitrogen and nitrogen compounds have been detected in interstellar space by astronomers using the Far Ultraviolet Spectroscopic Explorer. Molecular nitrogen is a major constituent of the Saturnian moon Titan's thick atmosphere, and occurs in trace amounts in other planetary atmospheres.

Nitrogen is present in all living organisms, in proteins, nucleic acids and other molecules. It typically makes up around 4% of the dry weight of plant matter, and around 3% of the weight of the human body. It is a large component of animal waste (for example, guano), usually in the form of urea, uric acid, ammonium compounds and derivatives of these nitrogenous products, which are essential nutrients for all plants that are unable to fix atmospheric nitrogen.

Nitrogen occurs naturally in a number of minerals, such as saltpetre (potassium nitrate), Chile saltpetre (sodium nitrate) and sal ammoniac (ammonium chloride). Most of these are relatively uncommon, partly because of the minerals' ready solubility in water.

Compounds

The main neutral hydride of nitrogen is ammonia (NH_3), although hydrazine (N_2H_4) is also commonly used. Ammonia is more basic than water by 6 orders of magnitude. In solution ammonia forms the ammonium ion (NH_4^+). Liquid ammonia (b.p. 240 K) is amphiprotic (displaying either Brønsted-Lowry acidic or basic character) and forms ammonium and the less common amide ions (NH_2^-); both amides and nitride (N^{3-}) salts are known, but decompose in water. Singly, doubly, triply and quadruply substituted alkyl compounds of ammonia are called amines (four substitutions, to form commercially and biologically important quaternary amines, results in a positively charged nitrogen, and thus a water-soluble, or at least amphiphilic, compound). Larger chains, rings and structures of nitrogen hydrides are also known, but are generally unstable. N_2^{2+} is another polyatomic cation as in hydrazine.

Other classes of nitrogen anions (negatively charged ions) are the poisonous azides (N_3^-), which are linear and isoelectronic to carbon dioxide, but which bind to important iron-containing enzymes in the body in a manner more resembling cyanide. Another molecule of the same structure is the colourless and relatively inert anesthetic gas Nitrous oxide (dinitrogen monoxide, N_2O), also known as laughing gas. This is one of a variety of nitrogen oxides that form a family often abbreviated as NOx. Nitric oxide (nitrogen monoxide, NO), is a natural free radical used in signal transduction in both plants and animals, for example in vasodilation by causing the smooth muscle of blood vessels to relax. The reddish and poisonous nitrogen dioxide NO_2 contains an unpaired electron and is an important component of smog. Nitrogen molecules containing unpaired electrons show an understandable tendency to dimerise (thus pairing the electrons), and are generally highly reactive. The corresponding acids are nitrous HNO_2 and nitric acid HNO_3, with the corresponding salts called nitrites and nitrates.

The higher oxides dinitrogen trioxide N_2O_3, dinitrogen tetroxide N_2O_4 (DTO) and dinitrogen pentoxide N_2O_5, are fairly unstable and explosive, a consequence of the chemical stability of N_2. DTO is one of the most important oxidisers of rocket fuels, used to oxidise hydrazine in the Titan rocket and in the recent NASA MESSENGER probe to Mercury. DTO is an intermediate in the manufacture of nitric acid HNO_3, one of the few acids stronger than hydronium and a fairly strong oxidising agent.

Nitrogen is notable for the range of explosively unstable compounds that it can produce. Nitrogen triiodide NI_3 is an extremely sensitive contact explosive. Nitrocellulose, produced by nitration of cellulose with nitric acid, is also known as guncotton. Nitroglycerin, made by nitration of glycerin, is the dangerously unstable explosive ingredient of dynamite. The comparatively stable, but more powerful explosive trinitrotoluene (TNT) is the standard explosive against which the power of nuclear explosions are measured.

Nitrogen can also be found in organic compounds. Common nitrogen functional groups include: amines, amides, nitro groups, imines, and enamines. The amount of nitrogen in a chemical substance can be determined by the Kjeldahl method.

Applications

Nitrogen gas is an industrial gas produced by the fractional distillation of liquid air, or by mechanical means using gaseous air (i.e. pressurised reverse osmosis membrane or Pressure swing adsorption). Commercial nitrogen is often a by-product of air-processing for industrial concentration of oxygen for steel-making and other purposes. When supplied compressed in cylinders it is often referred to as OFN (oxygen-free nitrogen).

Nitrogen gas has a wide variety of applications, including serving as an inert replacement for air where oxidation is undesirable:

- To preserve the freshness of packaged or bulk foods (by delaying rancidity and other forms of oxidative damage).
- In ordinary incandescent light bulbs as an inexpensive alternative to argon.
- On top of liquid explosives as a safety measure.
- The production of electronic parts such as transistors, diodes, and integrated circuits.
- Dried and pressurised, as a dielectric gas for high voltage equipment.
- The manufacturing of stainless steel
- Use in military aircraft fuel systems to reduce fire hazard.
- Filling automotive and aircraft tires due to its inertness and lack of moisture or oxidative qualities, as opposed to air, though this is not necessary for consumer automobiles.

Nitrogen molecules are less likely to escape from the inside of a tire compared with the traditional air mixture used. Air consists mostly of nitrogen and oxygen. Nitrogen molecules have a larger effective diameter than oxygen molecules and therefore diffuse through porous substances more slowly.

Nitrogen is commonly used during sample preparation procedures for chemical analysis. Specifically, it is used as a means of concentrating and reducing the volume of liquid samples. Directing a pressurised stream of nitrogen gas perpendicular to the surface of the liquid allows the solvent to evaporate while leaving the solute(s) and unevaporated solvent behind.

Nitrogen tanks are also replacing carbon dioxide as the main power source for paintball guns. The downside is that nitrogen must be kept at higher pressure than CO_2, making N_2 tanks heavier and more expensive.

Nitrogenated Beer

A further example of its versatility is its use as a preferred alternative to carbon dioxide to pressurise kegs of some beers, particularly stouts and British ales, due to the smaller bubbles it produces, which make the dispensed beer smoother and headier. A modern application of a pressure sensitive nitrogen capsule known commonly as a "widget" now allows nitrogen charged beers to be packaged in cans and bottles.

Liquid Nitrogen

Liquid nitrogen is a cryogenic liquid. At atmospheric pressure, it boils at "195.8 °C. When insulated in proper containers such as dewar flasks, it can be transported without much evaporative loss.

Like dry ice, the main use of liquid nitrogen is as a refrigerant. Among other things, it is used in the cryopreservation of blood, reproductive cells (sperm and egg), and other biological samples and materials. It is used in cold traps for certain laboratory equipment and to cool x-ray detectors. It has also been used to cool central processing units and other devices in computers which are overclocked, and which produce more heat than during normal operation.

Applications of Nitrogen Compounds

Molecular nitrogen (N_2) in the atmosphere is relatively non-reactive due to its strong bond, and N_2 plays an inert role in the human body, being neither produced or destroyed. In nature, nitrogen is converted into biologically (and industrially) useful compounds by lightning, and by some living organisms, notably certain bacteria (i.e. nitrogen fixing bacteria). Molecular nitrogen is released into the atmosphere in the process of decay, in dead plant and animal tissues.

The ability to combine or fix molecular nitrogen is a key feature of modern industrial chemistry, where nitrogen and natural gas are converted into ammonia via the Haber process. Ammonia, in turn, can be used directly (primarily as a fertilizer, and in the synthesis of nitrated fertilizers), or as a precursor of many other important materials including explosives, largely via the production of nitric acid by the Ostwald process.

The organic and inorganic salts of nitric acid have been important historically as convenient stores of chemical energy. They include important compounds such as potassium nitrate (or saltpeter used in gunpowder) and ammonium nitrate, an important fertilizer and explosive.

Various other nitrated organic compounds, such as nitroglycerin and trinitrotoluene, and nitrocellulose, are used as explosives and propellants for modern firearms. Nitric acid is used as an oxidising agent in liquid fuelled rockets. Hydrazine and hydrazine derivatives find use as rocket fuels and monopropellants.

In most of these compounds, the basic instability and tendency to burn or explode is derived from the fact that nitrogen is present as an oxide, and not as the far more stable nitrogen molecule (N_2) which is a product of the compounds' thermal decomposition. When nitrates burn or explode, the formation of the powerful triple bond in the N_2 which results, produces most of the energy of the reaction.

Nitrogen is a constituent of molecules in every major drug class in pharmacology and medicine. Nitrous oxide (N_2O) was discovered early in the 19th century to be a partial anaesthetic, though it was not used as a surgical anaesthetic until later. Called "laughing gas", it was found capable of inducing a state of social disinhibition resembling drunkenness. Other notable nitrogen-containing drugs are drugs derived from plant alkaloids, such as morphine (there exist many alkaloids known to have pharmacological effects; in some cases they appear natural chemical defences of plants against predation). Nitrogen containing drugs include all of the major classes of antibiotics, and organic nitrate drugs like nitroglycerin and nitroprusside which regulate blood pressure and heart action by mimicking the action of nitric oxide.

Biological Role

Nitrogen is an essential building block of both amino acids and nucleic acids, essential to life on Earth.

Elemental nitrogen in the atmosphere cannot be used directly by either plants or animals, and must converted to a reduced (or 'fixed') state in order to be useful for higher plants and animals. Precipitation often contains substantial quantities of ammonium and nitrate, thought to result from nitrogen fixation by lightning and other atmospheric electric phenomena. This was first proposed by Liebig in 1827 and later confirmed. However, because ammonium is preferentially retained by the forest canopy relative to atmospheric nitrate, most fixed nitrogen that reaches the soil surface under trees as nitrate. Soil nitrate is preferentially assimilated by these tree roots relative to soil ammonium.

Specific bacteria (e.g. *Rhizobium trifolium*) possess nitrogenase enzymes which can fix atmospheric nitrogen into a form (ammonium ion) that is chemically useful to higher organisms. This process requires a large amount of energy and anoxic conditions. Such bacteria may live freely in soil (e.g. *Azotobacter*) but normally exist in a symbiotic relationship in the root nodules of leguminous plants (e.g. clover, *Trifolium*, or soya bean plant, *Glycine max*). Nitrogen-fixing bacteria are also symbiotic with a number of unrelated plant species such as alders (*Alnus*) spp., lichens (*Casuarina*), *Myrica*, liverworts, and *Gunnera*.

As part of the symbiotic relationship, the plant converts the 'fixed' ammonium ion to nitrogen oxides and amino acids to form proteins and other molecules, (e.g. alkaloids). In return for the 'fixed' nitrogen, the plant secretes sugars to the symbiotic bacteria.

Some plants are able to assimilate nitrogen directly in the form of nitrates which may be present in soil from natural mineral deposits, artificial fertilizers, animal waste, or organic decay (as the product of bacteria, but not bacteria specifically associated with the plant). Nitrates absorbed in this fashion are converted to nitrites by the enzyme *nitrate* reductase, and then converted to ammonia by another enzyme called *nitrite* reductase.

Nitrogen compounds are basic building blocks in animal biology as well. Animals use nitrogen-containing amino acids from plant sources, as starting materials for all nitrogen-compound animal biochemistry, including the manufacture of proteins and nucleic acids. Plant-feeding insects are dependent on nitrogen in their diet, such that varying the amount of nitrogen fertilizer applied to a plant can affect the reproduction rate of insects feeding on fertilized plants.

Soluble nitrate is an important limiting factor in the growth of certain bacteria in ocean waters. In many places in the world, artificial fertilizers applied to crop-lands to increase yields result in run-off delivery of soluble nitrogen to oceans at river mouths. This process can result in eutrophication of the water, as nitrogen-driven bacterial growth depletes water oxygen to the point that all higher organisms die. Well-known "dead zone" areas in the US Gulf Coast and the Black Sea are due to this important polluting process.

Many saltwater fish manufacture large amounts of trimethylamine oxide to protect them from the high osmotic effects of their environment (conversion of this compound to dimethylamine is responsible for the early odour in not fresh saltwater fish. In animals, free radical nitric oxide (*NO*) (derived from an amino acid), serves as an important regulatory molecule for circulation.

Animal metabolism of *NO* results in production of nitrite. Animal metabolism of nitrogen in proteins generally results in excretion of urea, while animal metabolism of nucleic acids results in excretion of urea and uric acid. The characteristic odour of animal flesh decay is caused by the creation of long-chain, nitrogen-containing amines, such as putrescine and cadaverine.

Decay of organisms and their waste products may produce small amounts of nitrate, but most decay eventually returns nitrogen content to the atmosphere, as molecular nitrogen. The circulation of nitrogen from atmosphere to organic compounds and back is referred to as the nitrogen cycle.

Safety

Rapid release of nitrogen gas into an enclosed space can displace oxygen, and therefore represents an asphyxiation hazard. This may happen with few warning symptoms, since the human carotid body is a relatively slow and a poor low-oxygen (hypoxia) sensing system.

An example occurred shortly before the launch of the first Space Shuttle mission in 1981, when two technicians lost consciousness and died after they walked into a space located in the Shuttle's Mobile Launcher Platform that was pressurised with pure nitrogen as a precaution against fire. The technicians would have been able to exit the room if they had experienced early symptoms from nitrogen-breathing.

When inhaled at high partial pressures (more than about 4 bar, encountered at depths below about 30 m in scuba diving) nitrogen begins to act as an anaesthetic agent. It can cause nitrogen narcosis, a temporary semi-anaesthetised state of mental impairment similar to that caused by nitrous oxide.

Nitrogen also dissolves in the bloodstream and body fats. Rapid decompression (particularly in the case of divers ascending too quickly, or astronauts decompressing too quickly from cabin pressure to spacesuit pressure) can lead to a potentially fatal condition called decompression sickness (formerly known as caisson sickness or more commonly, the "bends"), when nitrogen bubbles form in the bloodstream, nerves, joints, and other sensitive or vital areas.

Other "inert" gases (those gases other than carbon dioxide and oxygen) cause the same effects from bubbles composed of them, so replacement of nitrogen in breathing gases may prevent nitrogen narcosis, but does not prevent decompression sickness.

Direct skin contact with liquid nitrogen will eventually cause severe frostbite (cryogenic burns). This may happen almost instantly on contact, depending on the form of liquid nitrogen. Bulk liquid nitrogen causes less rapid freezing than a spray of nitrogen mist (such as is used to freeze certain skin growths in the practice of dermatclogy).

The extra surface area provided by nitrogen-soaked materials is also important, with soaked clothing or cotton causing far more rapid damage than a spill of direct liquid to skin. Full "contact" between naked skin and large droplets or pools of undisturbed liquid nitrogen may be prevented for a few seconds by a layer of insulating gas from the Leidenfrost effect. However, liquid nitrogen applied to skin in mists, and on fabrics, bypasses this effect.

15

The Nutrition

Nutrition (also called nourishment or aliment) is the provision, to cells and organisms, of the materials necessary (in the form of food) to support life. Many common health problems can be prevented or alleviated with a healthy diet.

The *diet* of an organism refers to what it eats. Dietitians are health professionals who specialise in human nutrition, meal planning, economics, preparation, and so on. They are trained to provide safe, evidence-based dietary advice and management to individuals (in health and disease), as well as to institutions.

Poor diet can have an injurious impact on health, causing deficiency diseases such as scurvy, beriberi, and kwashiorkor; health-threatening conditions like obesity and metabolic syndrome, and such common chronic systemic diseases as cardiovascular disease, diabetes, and osteoporosis.

Nutritional science investigates the metabolic and physiological responses of the body to diet. With advances in the fields of molecular biology, biochemistry, and genetics, the study of nutrition is increasingly concerned with metabolism and metabolic pathways: the sequences of biochemical steps through which substances in living things change from one form to another.

The human body contains chemical compounds, such as water, carbohydrates (sugar, starch, and fibre), amino acids (in proteins), fatty acids (in lipids), and nucleic acids (DNA and RNA). These compounds in turn consist of elements such as carbon, hydrogen, oxygen, nitrogen, phosphorus, calcium, iron, zinc, magnesium, manganese, and so on. All of these chemical compounds and elements occur in various forms and combinations (e.g. hormones, vitamins, phospholipids, hydroxyapatite), both in the human body and in the plant and animal organisms that humans eat.

The human body consists of elements and compounds ingested, digested, absorbed, and circulated through the bloodstream to feed the cells of the body. Except in the unborn fetus, the digestive system is the first system involved. In a typical adult, about seven litres of digestive juices enter the lumen of the digestive tract. These break chemical bonds in ingested molecules, and modulate their conformations and energy states. Though some molecules are absorbed into the bloodstream unchanged, digestive processes release them from the matrix of foods. Unabsorbed matter, along with some waste products of metabolism, is eliminated from the body in the feces.

Studies of nutritional status must take into account the state of the body before and after experiments, as well as the chemical composition of the whole diet and of all material excreted and eliminated from the body (in urine and feces). Comparing the food to the waste can help determine the specific compounds and elements absorbed and metabolised in the body. The effects of nutrients may only be discernible over an extended period, during which all food and waste must be analysed. The number of variables involved in such experiments is high, making nutritional studies time-consuming and expensive, which explains why the science of human nutrition is still slowly evolving.

In general, eating a wide variety of fresh, whole (unprocessed), foods has proven favourable compared to monotonous diets based on processed foods. In particular, the consumption of whole-plant foods slows digestion and allows better absorption, and a more favourable balance of essential nutrients per Calorie, resulting in better management of cell growth, maintenance, and mitosis (cell division), as well as better regulation of appetite and blood sugar. Regularly scheduled meals (every few hours) have also proven more wholesome than infrequent or haphazard ones.

The Nutrients

There are seven major classes of nutrients: carbohydrates, fats, fibre, minerals, protein, vitamins, and water.

These nutrient classes can be categorised as either macronutrients (needed in relatively large amounts) or micronutrients (needed in smaller quantities). The macronutrients are carbohydrates, fats, fibre, proteins, and water. The micronutrients are minerals and vitamins.

The macronutrients (excluding fibre and water) provide energy, which is measured in Joules or kilocalories (often called "Calories" and written with a capital *C* to distinguish them from gram calories). Carbohydrates and proteins provide 17 kJ (4 kcal) of energy per gram, while fats provide 37 kJ (9 kcal) per gram. Vitamins, minerals, fibre, and water do not provide energy, but are necessary for other reasons.

Molecules of carbohydrates and fats consist of carbon, hydrogen, and oxygen atoms. Carbohydrates range from simple monosaccharides (glucose, fructose, galactose) to complex polysaccharides (starch). Fats are triglycerides, made of various fatty acid monomers bound to glycerol.

Some fatty acids, but not all, are essential in the diet: they cannot be synthesised in the body. Protein molecules contain nitrogen atoms in addition to the elements of carbohydrates and fats. The nitrogen-containing monomers of protein are amino acids, and they include some essential amino acids. They fulfil many roles other than energy metabolism; and when they are used as fuel, getting rid of the nitrogen places a burden on the kidneys.

Other micronutrients include antioxidants and phytochemicals.

Most foods contain a mix of some or all of the nutrient classes. Some nutrients are required regularly, while others are needed only occasionally. Poor health can be caused by an imbalance of nutrients, whether an excess or a deficiency.

Carbohydrates

Carbohydrates may be classified as monosaccharides, disaccharides, or polysaccharides by the number of monomer (sugar) units they contain. They constitute a large proportion of foods such as rice, noodles, bread, and other grain-based products. Monosaccharides contain one sugar unit, disaccharides two, and polysaccharides three or more. Polysaccharides are often referred to as *complex* carbohydrates because they are long chains of sugar units, whereas monosaccharides and disaccharides are simpler. The difference is important: complex carbohydrates take longer to digest and absorb since their sugar units are processed one-by-one off the ends of the chains. The spike in blood-sugar levels after ingestion of simple sugars is thought to be involved in causing heart and vascular disease. Simple sugars form a greater part of modern diets, leading to more cardiovascular disease in populations. Simple carbohydrates are absorbed quickly, and therefore raise blood-sugar levels more rapidly.

Fats

A molecule of fat consists of several fatty acids (containing long chains of carbon and hydrogen atoms), bonded to a glycerol. They are typically found as triglycerides (three fatty acids attached to one glycerol backbone). Fats may be classified as saturated or unsaturated. Saturated fats have all of the carbon atoms in their fatty acid chains bonded to hydrogen atoms, whereas unsaturated fats have some of these carbon atoms double-bonded, so their molecules have relatively few hydrogen atoms. Unsaturated fats may be further classified as monounsaturated (one double-bond) or polyunsaturated (many double-bonds). Trans fats are a type of unsaturated fat with *trans*-isomer fatty acid, typically created in an industrial process called (partial) hydrogenation.

Many studies have shown that unsaturated fats, particularly monounsaturated fats, are best in the human diet. Saturated fats, typically from animal sources, are next, while trans fats are to be avoided. Saturated and trans fats are typically solid at room temperature (such as butter or lard), while unsaturated fats are typically liquids (such as olive oil or flaxseed oil). Trans fats are very rare in nature, but have properties useful in the food processing industry.

Essential Fatty Acids

Most fatty acids are non-essential, meaning the body can produce them as needed. However, in humans at least two fatty acids are essential and must be included in the diet. An appropriate balance of essential fatty acids—omega-3 and omega-6 fatty acids—is important for health. Both of these "omega" long-chain polyunsaturated fatty acids are substrates for a class of eicosanoids known as prostaglandins, which have roles throughout the human body. They are hormones, in some respects. The omega-3 eicosapentaenoic acid (EPA), which can be made in the human body from the omega-3 essential fatty acid alpha-linolenic acid (LNA), or taken in through marine food sources, serves as a building block for series 3 prostaglandins (e.g. weakly inflammatory PGE3).

The omega-6 dihomo-gamma-linolenic acid (DGLA) serves as a building block for series 1 prostaglandins (e.g. anti-inflammatory PGE1), whereas arachidonic acid (AA) serves as a building block for series 2 prostaglandins (e.g. pro-inflammatory PGE 2). Both DGLA and AA can be made from the omega-6 linoleic acid (LA) in the human body, or can be taken in directly through food. An appropriately balanced intake of omega-3 and omega-6 partly determines the relative production of different prostaglandins: one reason a balance between omega-3 and omega-6 is important for cardiovascular health. In industrialised societies, people typically consume large amounts of processed vegetable oils, which have reduced amounts of the essential fatty acids along with too much of omega-6 fatty acids relative to omega-3 fatty acids.

The conversion rate of omega-6 DGLA to AA largely determines the production of the prostaglandins PGE1 and PGE2. Omega-3 EPA prevents AA from being released from membranes, thereby skewing prostaglandin balance away from pro-inflammatory PGE2 (made from AA) towards anti-inflammatory PGE1 (made from DGLA). Moreover, the conversion (desaturation) of DGLA to AA is controlled by the enzyme delta-5-desaturase, which in turn is controlled by hormones such as insulin (up-regulation) and glucagon (down-regulation). The amount and type of carbohydrates consumed, along with some types of amino acid, can influence processes involving insulin, glucagon, and other hormones; therefore the ratio of omega-3 versus omega-6 has wide effects on general health, and specific effects on immune function and inflammation, and mitosis (i.e. cell division).

Good sources of essential fatty acids include most vegetables, nuts, seeds, and marine oils, Some of the best sources are fish, flax seed oils, soy beans, pumpkin seeds, sunflower seeds, and walnuts.

Fibre

Fibre is a carbohydrate (or a polysaccharide) that is incompletely absorbed in humans and in some other animals. Like all carbohydrates, when it is metabolised it can produce four Calories (kilocalories) of energy per gram. But in most circumstances it accounts for less than that because of its limited absorption. Dietary fibre consists mainly of cellulose, a large carbohydrate polymer that is indigestible because humans do not have the required enzymes.

There are two subcategories: soluble and insoluble fibre. Whole grains, fruits (especially plums, prunes, and figs), and vegetables are rich in dietary fibre. Fibre is important to digestive health and is thought to reduce the risk of colon cancer. It can help in alleviating both constipation and diarrhoea. Fibre provides bulk to the intestinal contents, and insoluble fibre stimulates peristalsis: the rhythmic muscular contractions passing along the digestive tract. Some soluble fibres produce a solution of high viscosity: a gel, which slows the movement of food through the intestines. Fibre, especially from whole grains, may help lessen insulin spikes and reduce the risk of diabetes [type 2].

Protein

Proteins are the basis of many animal body structures (e.g. muscles, skin, and hair). Each molecule is composed of amino acids which are characterised by inclusion of nitrogen and sometimes sulphur. The body requires amino acids to produce new proteins (protein retention) and to replace damaged proteins (maintenance). Excess amino acids are discarded, typically in the urine. For all animals, some amino acids are *essential* (an animal cannot produce them internally) and some are *non-essential* (the animal can produce them from other nitrogen-containing compounds).

About twenty amino acids are found in the human body, and about ten of these are essential, and therefore must be included in the diet. A diet that contains adequate amounts of amino acids (especially those that are essential) is particularly important when there is greater need: in early development and maturation, pregnancy, lactation, or injury.

A *complete* protein source contains all the essential amino acids; an *incomplete* protein source lacks one or more of the essential amino acids. It is possible to combine two incomplete protein sources (e.g. rice and beans) to make a complete protein source. Sources of dietary protein include meats, tofu and other soy-products, eggs, grains, legumes, and dairy products such as milk and cheese. A few amino acids from protein can be converted into glucose and used for fuel through a process called gluconeogenesis. The amino acids remaining after such conversion are discarded.

Minerals

Dietary minerals are the chemical elements required by living organisms, other than the four elements carbon, hydrogen, nitrogen, and oxygen that are present in common organic molecules. The term "mineral" is archaic, since the intent is to describe simply the less common elements in the diet: heavier than the four just mentioned; including several metals; and often occurring as ions in the body. Some dietitians recommend that these be supplied from foods in which they occur naturally, or at least as complex compounds, or sometimes even from natural inorganic sources (such as calcium carbonate from ground oyster shells). On the other hand, minerals are often artificially added to the diet as supplements, the most famous being iodine in iodised salt.

Macrominerals

Many elements are essential in quantity; also called "bulk minerals". Some are structural, but many play a role as electrolytes. Elements with recommended dietary allowance (RDA) greater than 200 mg/day are the following, in alphabetical order (with informal or folk-medicine perspectives in parentheses):

- Calcium, a common electrolyte, but also structural (for muscle and digestive system health, builds bone, neutralises acidity, clears toxins, helps blood stream).
- Chlorine as chloride ions; very common electrolyte.
- Magnesium, required for processing ATP and related reactions (builds bone, causes strong peristalsis, increases flexibility, increases alkalinity).
- Phosphorus, required component of bones; essential for energy processing.
- Potassium, a very common electrolyte (heart and nerve health).
- Sodium, a very common electrolyte; not generally found in dietary supplements, despite being needed in large quantities, because the ion is very common in food: typically as sodium chloride, or common salt.
- Sulphur for three essential amino acids and therefore many proteins (skin, hair, nails, liver, and pancreas).

Trace Minerals

Many elements are required in trace amounts, usually because they play a catalytic role in enzymes. Some trace mineral elements (RDA < 200 mg/day) are, in alphabetical order:

- Cobalt required for biosynthesis of vitamin B12 family of coenzymes.
- Copper required component of many redox enzymes, including cytochrome c oxidase.
- Chromium required for sugar metabolism.
- Iodine required for the biosynthesis of thyroxin; needed in larger quantities than others in this list, and sometimes classified with the macrominerals.
- Iron required for many enzymes, and for haemoglobin and some other proteins.
- Manganese (processing of oxygen).
- Molybdenum required for xanthine oxidase and related oxidases.
- Nickel present in urease.
- Selenium required for peroxidase (antioxidant proteins).
- Vanadium (Speculative: there is no established RDA for vanadium. No specific biochemical function has been identified for it in humans, although vanadium is found in lower organisms).
- Zinc required for several enzymes such as carboxypeptidase, liver alcohol dehydrogenase, carbonic anhydrase.

Vitamins

As with the minerals, twelve vitamins are recognised as essential nutrients, necessary in the diet for good health. (Vitamin D is the exception: it can alternatively be synthesised in the skin, in the presence of UVB radiation.) Certain vitamin-like compounds that are recommended in the diet, such as carnitine, are indispensable for survival and health; but these are not strictly "essential" because the human body has some capacity to produce them from other compounds. Moreover, thousands of different phytochemicals have recently been discovered in food (particularly in fresh vegetables), which may have desirable properties including antioxidant activity. Other essential nutrients not classed as vitamins include essential amino acids, choline, essential fatty acids, and the minerals.

Vitamin deficiencies may result in disease conditions: goitre, scurvy, osteoporosis, impaired immune system, disorders of cell metabolism, certain forms of cancer, symptoms of premature aging, and poor psychological health (including eating disorders), among many others. Excess of some vitamins is also dangerous to health (notably vitamin A); and deficiency or excess of minerals can also have serious health consequences.

Water

About 70% of the non-fat mass of the human body is made of water. To function properly, the body requires between one and seven litres of water per day to avoid dehydration; the precise amount depends on the level of activity, temperature, humidity, and other factors. With physical exertion and heat exposure, water loss will increase and daily fluid needs may increase as well.

It is not clear how much water intake is needed by healthy people, although some experts assert that 8–10 glasses of water (approximately 2 litres) daily is the minimum to maintain proper hydration. The notion that a person should consume eight glasses of water per day cannot be traced back to a scientific source. The effect of water intake on weight loss and on constipation is also still unclear. Original recommendation for water intake in 1945 by the Food and Nutrition Board of the National Research Council read: "An ordinary standard for diverse persons is 1 milliliter for each calorie of food. Most of this quantity is contained in prepared foods." The latest dietary reference intake report by the United States National Research Council in general recommended (including food sources):

2.7 litres of water total for women and 3.7 litres for men. Specifically, pregnant and breastfeeding women need additional fluids to stay hydrated. According to the Institute of Medicine—who recommend that, on average, women consume 2.2 litres and men 3.0 litres—this is recommended to be 2.4 litres (approx. 9 cups) for pregnant women and 3 litres (approx. 12.5 cups) for breastfeeding women since an especially large amount of fluid is lost during nursing.

For those who have healthy kidneys, it is rather difficult to drink too much water, but (especially in warm humid weather and while exercising) it is dangerous to drink too little. People can drink far more water than necessary while exercising, however, putting them at risk of water intoxication, which can be fatal. In particular large amounts of de-ionised water are dangerous.

Normally, about 20 percent of water intake comes in food, while the rest comes from drinking water and assorted beverages (caffeinated included). Water is excreted from the body in multiple forms; including urine and feces, sweating, and by water vapour in the exhaled breath.

Other Nutrients

Other micronutrients include antioxidants and phytochemicals. These substances are generally more recent discoveries which: have not yet been recognised as vitamins; are still under investigation; or contribute to health but are not necessary for life. Phytochemicals may act as antioxidants, but not all phytochemicals are antioxidants.

16

Fertility

Soil is the characteristics that supports life. In particular, the term is used to describe agriculture and garden life.

Fertile soil has the following properties:

- It is rich in nutrients necessary for basic plant nutrition, including nitrogen, phosphorus and potassium.
- It contains sufficient minerals (trace elements) for plant nutrition, including boron, chlorine, cobalt, copper, iron, manganese, magnesium, molybdenum, sulphur, and zinc.
- It contains soil organic matter that improves soil structure and soil moisture retention.
- Soil pH is in the range 6.0 to 6.8.
- Good soil structure, creating well drained soil.
- A range of microorganisms that support plant growth.
- It often contains large amounts of topsoil.

In lands used for agriculture and other human activities, fertile soil typically arises from the use of soil conservation practices.

Soil Fertilization

Nitrogen is the element in the soil that is most often lacking. Phosphorus and potassium are also needed in substantial amounts. For this reason these three elements are always included in commercial fertilizers and the content of each of these items is included on the bags of fertilizer. For example a 10-10-15 fertilizer has 10 percent nitrogen, 10 percent (P_2O_5) available phosphorus and 15 percent (K_2O) water soluble potassium. Inorganic fertilizers are generally less expensive and have higher concentrations of nutrients than organic fertilizers.

Some have criticised the use of inorganic fertilizers claiming that the water-soluble nitrogen doesn't provide for the long-term needs of the plant and creates water pollution. Slow-release fertilizer, however, is not soluble and eliminates the biggest negative of fertilization fertilizer burn. Additionally, most soluble fertilizers are coated, such as sulphur-coated urea.

In 2008, the cost of phosphorus as fertilizer more than doubled while the price of rock phosphate as base commodity rose 8-fold, recently the term peak phosphorus has been coined, due to the limited occurrence of rock phosphate in the world.

Soil can be revitalised through physical means such as soil steaming as well. Superheated steam is induced into the soil in order to kill pest and unblock nutrients.

Light and CO_2 Limitations

Photosynthesis is the process whereby plants use light energy to drive chemical reactions which convert CO_2 into sugars. As such, all plants require access to both light and carbon dioxide to produce energy, grow and reproduce.

While typically limited by nitrogen, phosphorus and potassium, low levels of carbon dioxide can also act as a limiting factor on plant growth. Peer reviewed and published scientific studies have shown that increasing CO_2 is highly effective at promoting plant growth up to levels over 300 ppm. Further increases in CO_2 can, to a very small degree, continue to increase net photosynthetic output (Chapin *et al.*, 2002 - Principles of Terrestrial Ecosystem Ecology).

Since higher levels of CO_2 have only a minimal impact on photosynthetic output at present levels (presently around 380 ppm and increasing), we should not consider plant growth to be limited by carbon dioxide. Other biochemical limitations, such as soil organic content, nitrogen in the soil, phosphorus and potassium, are far more often in short supply. As such, neither commercial nor scientific communities look to air fertilization as an effective or economic method of increasing production in agriculture or natural ecosystems. Furthermore, since microbial decomposition occurs faster under warmer temperatures, higher levels of CO_2 (which is one of the causes of unusually fast climate change) should be expected to increase the rate at which nutrients are leached out of soils and may have a negative impact on soil fertility.

Soil Depletion

Soil depletion occurs when the components which contribute to fertility are removed and not replaced, and the conditions which support soil fertility are not maintained. This leads to poor crop yields. In agriculture, depletion can be due to excessively intense cultivation and inadequate soil management.

One of the most widespread occurrences of soil depletion as of 2008 is in tropical zones where nutrient content of soils is low. The combined effects of growing population densities, large-scale industrial logging, slash-and-burn agriculture and ranching, and other factors, have in some places depleted soils through rapid and almost total nutrient removal.

Topsoil depletion is when the nutrient rich organic topsoil that takes hundreds to thousands of years to build up under natural conditions is eroded or depleted of its original organic material. Historically, many past civilizations collapses can be attributed to the depletion of the topsoil. Since the beginning of agricultural production in the Great Plains of North America in the 1880s about one half of its topsoil has disappeared.

Depletion may occur through a variety of other effects, including overtillage which damages soil structure, overuse of inputs such as synthetic fertilizers and herbicides, which leave residues and build-ups that inhibit microorganisms, and salinisation of soil.

Plaggen Soil

Plaggen is a type of soil created in Europe in the Middle Ages, as a result of so-called 'plaggen cultivation', created by cutting turves of peat from an outfield area, and then using them as bedding for cattle; the slurry-soaked bedding was later spread on the arable fields as fertilizer. Over time, this created a very rich agricultural soil which could be over 1 m in depth—unlike our modern arable soils, which tend to be just 30 cm deep.

In Orkney these soils were created already in the 12th to 13th Centuries, and on some islands in Shetland these methods continued to be used until the 1960s.

Shifting Cultivation

Shifting cultivation is an agricultural system in which plots of land are cultivated temporarily, then abandoned. This system often involves clearing of a piece of land fallowed by several years of wood harvesting or farming, until the soil loses fertility. Once the land becomes inadequate for crop production, it is left to be reclaimed by natural vegetation, or sometimes converted to a different long-term cyclical farming practice. And ecological consequences are often deleterious, but are not as severe provided new forests are not invaded. Of these cultivators, many use a practice of slash-and-burn as one element of their farming cycle. Others employ land clearing without any burning, and some cultivators are purely migratory and do not use any cyclical method on a given plot. Sometimes no slashing at all is needed where regrowth is purely of grasses, an outcome not uncommon when soils are near exhaustion and need to lie fallow.

The Political Ecology of Shifting Cultivation

Shifting cultivation is a form of agriculture in which the cultivated or cropped area is shifted regularly to allow soil properties to recover under conditions of natural successive stages of re-growth. In a shifting cultivation system, at any particular point in time a minority of 'fields' are in cultivation and a majority are in various stages of natural re-growth. Over time, fields are cultivated for a relatively short time, and allowed to recover, or are followed, for a relatively long time. Eventually a previously cultivated field will be cleared of the natural vegetation and planted in crops again. Fields in established and stable shifting cultivation systems are cultivated and fallowed cyclically.

Fallow fields are not unproductive. During the fallow period, shifting cultivators use the successive vegetation species widely for timber for fencing and construction, firewood, thatching, ropes, clothing, tools, carrying devices and medicines. It is common for fruit and nut trees in fallows to be planted in fallow fields to the extent that parts of some fallows are in fact orchards. Soil-enhancing shrub or tree species may be planted or protected from slashing or burning in fallows. Many of these species have been shown to fix nitrogen. Fallows commonly contain plants that attract birds and animals and are important for hunting. But perhaps most importantly, tree fallows protect soil against physical erosion and draw nutrients to the surface from deep in the soil profile.

The relationship between the time the land is cultivated and the time it is fallowed are critical to the stability of shifting cultivation systems. These parameters determine whether or not the shifting cultivation system as a whole suffers a net loss of nutrients over time. A system in which there is a net loss of nutrients with each cycle will eventually lead to a degradation of resources unless actions are taken to arrest the losses. In some cases soil can be irreversibly exhausted (including erosion as well as nutrient loss) in less than a decade.

The longer a field is cropped, the greater the loss of soil organic matter, the reduction in the cation-exchange-capacity and in nitrogen and phosphorus, the greater the increase in acidity, the more likely soil porosity and infiltration capacity is reduced and the greater the loss of seeds of naturally occurring plant species from soil seed banks. In a stable shifting cultivation system, the fallow is long enough for the natural vegetation to recover to the state that it was in before it was cleared, and for the soil to recover to the condition it was in before cropping began. During fallow periods soil temperatures are lower, wind and water erosion is much reduced, nutrient cycling becomes closed again, nutrients are extracted from the subsoil, soil fauna increases, acidity is reduced, soil structure, texture and moisture characteristics improve and seed banks are replenished.

No universal optimum relationship exists between the length of the cropping period and the length of the fallow period. In favourable agricultural environments, cropping periods can be longer and fallow periods shorter, than in less favourable agricultural environments. In favourable environments soil conditions at the beginning of a cropping cycle will be better and fallow successional stages will proceed faster. Nevertheless, even in the most favourable environments, it is likely that if the cropping period is extended beyond a certain point, the fallow conditions required for an adequate recovery of soils and vegetation will be jeopardised.

If the fallow period is shortened there will be less time in which the soil recovery processes and vegetation successions can take place. The length of fallow period required to prevent net loss of nutrients will again depend on the quality of the environment, which will in turn, determine the rate at which recovery occurs. But sooner or later, if the fallow period continues to be reduced, an observable change will occur in the fallow vegetation. Secondary forest may be reduced to shorter, thinner stemmed, fewer, woody bush or jungle species, bush may be reduced to scrub and tall grasses and scrub and tall grasses may be reduced to short grasses. Less directly observable, but nevertheless critical changes will also be occurring in the soil. Changes in environmental conditions that happen subsequent to either a lengthening of the cropping period or a shortening of the fallow period often result in a fall in crop yields. It is not difficult to perceive how a shifting cultivation system, once destabilised, can proceed into a vicious circle of declining yields and shortening fallows or lengthening cropping periods, which in turn lead to further degradation of environmental conditions.

The secondary forests created by shifting cultivation are commonly richer in plant and animal resources useful to humans than primary forests, even though they are much less bio-diverse. Shifting cultivators view the forest as an agricultural landscape of fields at various stages in a regular cycle. People unused to living in forests cannot see the fields for the trees. Rather they perceive an apparently chaotic landscape in which trees are cut and burned randomly and so they characterise shifting cultivation as ephemeral or 'pre-agricultural', as 'primitive' and as a stage to be progressed beyond. Shifting agriculture is none of these things. Stable shifting cultivation systems are highly variable, closely adapted to micro-environments and are carefully managed by farmers during both the cropping and fallow stages. Shifting cultivators may possess a highly developed knowledge and understanding of their local environments and of the crops and native plant species they exploit. Complex and highly adaptive land tenure systems sometimes exist under shifting cultivation. Introduced crops for food and as cash have been skilfully integrated into some shifting cultivation systems.

Stereotypes: Primitive, Backward, Wasteful, Unproductive

Shifting cultivation systems are perceived both by numerous scientists as well as the general public, as primitive, backward, wasteful, unproductive, exploitative and the cause of widespread environmental degradation. Shifting cultivators are blamed for the destruction of much of the world's tropical forests, land degradation, atmospheric pollution and global climatic change. While contemporary manifestations of these attitudes towards shifting cultivators are often political and reflect competition between land occupying farmers, migrant settlers, loggers and international capital seeking access to tropical forests, they can sometimes be traced to the late-19th and early-20th century, to European colonial administrations in tropical and subtropical South Asia, South East Asia and South America. Indigenous occupants were then characterised by the colonisers as primitive and hence their agriculture systems, which were commonly a form of shifting cultivation were also viewed as primitive. In addition, there was a fundamental misunderstanding that shifting cultivators selected sites for cropping at random, thus destroying forests.

These attitudes may have prevented observers of tropical environments from understanding the social and economic conditions underlying shifting cultivation. It is ironic that in a major pioneering work in the recognition of the extent of human impacts upon the Earth published in the 1950s (Thomas 1956) a chapter on agriculture in the tropics states "No matter where we go we find primitive agriculture was carried on at the expense of forest" and makes a distinction between "primitive tropical horticulture" and "agriculture" as practiced by "higher cultures" (Bartlett 1956). All shifting cultivation systems are destructive of resources. In the same book another chapter describes massive loss of the forests in Europe prior to the beginning of the 20th century and details the social and economic conditions, other than shifting cultivation, that brought about their destruction (Darby 1956). Some of the conditions responsible for the deforestation of Europe are similar to those causing deforestation today.

Shifting Cultivation in Europe

Shifting cultivation was still being practiced as a viable and stable form of agriculture in many parts of Europe and west into Siberia at the end of the 19th century and in some places well into the 20th century. In the Ruhr in the late-1860s a forest-field rotation system known as Reutbergwirtschaft was using a 16 year cycle of clearing, cropping and fallowing with trees to produce bark for tanneries, wood for charcoal and rye for flour (Darby 1956, 200). Swidden farming was practiced in Siberia at least until the 1930s, using specially selected varieties of "swidden-rye" (Steensberg 1993, 98).

In Eastern Europe and Northern Russia the main swidden crops were turnips, barley, flax, rye, wheat, oats, radishes and millet. Cropping periods were usually one year, but were extended to two or three years on very favourable soils. Fallow periods were between 20 and 40 years (Linnard 1970, 195). In Finland in 1949, Steensberg (1993, 111) observed the clearing and burning of a 60,000 square metre swidden 440 km north of Helsinki. Birch and pine trees had been cleared over a period of a year and the logs sold for cash. A fallow of alder (Alnus) was encouraged to improve soil conditions. After the burn, turnip was sown for sale and for cattle feed. Shifting cultivation was disappearing in this part of Finland because of a loss of agricultural labour to the industries of the towns. Steensberg (1993, 110-152) provides eye-witness descriptions of shifting cultivation being practiced in Sweden in the 1900s, and in Estonia, Poland, the Caucasus, Serbia, Bosnia, Hungary, Switzerland, Austria and Germany in the 1930s to the 1950s.

That these agricultural practices survived from the Neolithic into the middle of the 20th century amidst the sweeping changes that occurred in Europe over that period, suggests they were adaptive and in themselves, were not massively destructive of the environments in which they were practiced. This raises the question: if shifting cultivation did not lead to the disappearance of European forests, what did?

The earliest written accounts of forest destruction in Southern Europe begin around 1000 BC in the histories of Homer, Thucydides and Plato and in Strabo's Geography. Forests were exploited for ship building, and urban development, the manufacture of casks, pitch and charcoal, as well as being cleared for agriculture. The intensification of trade and as a result of warfare, increased the demand for ships which were manufactured completely from forest products. Although goat herding is singled out as an important cause of environmental degradation, a more important cause of forest destruction was the practice in some places of granting ownership rights to those who clear felled forests and brought the land into permanent cultivation. Evidence that circumstances other than agriculture were the major causes for forest destruction was the recovery of tree cover in many parts of the Roman Empire from 400 BC to around 500 AD following the collapse of Roman economy and industry. Darby observes that by 400 AD "land that had once been tilled became derelict and overgrown" and quotes Lactantius who wrote that in many places "cultivated land became forest". (Darby 1956, 186). The other major cause of forest destruction in the Mediterranean environment with its hot dry summers were wild fires that became more common following human interference in the forests.

In Central and Northern Europe the use of stone tools and fire in agriculture is well established in the palynological and archaeological record from the Neolithic. Here, just as in Southern Europe, the demands of more intensive agriculture and the invention of the plough, trading, mining and smelting, tanning, building and construction in the growing towns and constant warfare, including the demands of naval shipbuilding, were more important forces behind the destruction of the forests than was shifting cultivation.

By the Middle Ages in Europe, large areas of forest were being cleared and converted into arable land in association with the development of feudal tenurial practices. From the 1500s to the 1700s the demands of iron smelters for charcoal, increasing industrial developments and the discovery and expansion of colonial empires as well as incessant warfare that increased the demand for shipping to levels never previously reached, all combined to deforest Europe.

With the loss of the forest, so shifting cultivation became restricted to the peripheral places of Europe, where permanent agriculture was uneconomic, transport costs constrained logging or terrain prevented the use of draught animals or tractors. It has disappeared from even these refuges since 1945, as agriculture has become increasingly capital intensive, rural areas have become depopulated and the remanent European forests themselves have been revalued economically and socially.

Simple Societies, Shifting Cultivation and Environmental Change

The forests of Europe were destroyed by the seemingly inexorable 'advances' of civilization, industrialisation and warfare, the sort of 'advances' that many of those who criticised shifting cultivators in the 19th century thought were desirable and indicative of "higher cultures". The same sort of processes are leading to the destruction of tropical forests in the last decade of the 20th century. So 'advances' in civilization, now known as 'development', have not resolved these problems. The problems are located not in the practice of a particular form of agriculture, but within the fundamental relationships that human societies have with their environments. In complex developed economies these relationships become very elaborate and are difficult to comprehend. However, in simple economies, where agriculture is the major source of wealth creation, they can be easier to understand.

A growing body of archaeological and palynological evidence finds that simple human societies brought about extensive changes to their environments before the establishment of any sort of state, feudal or capitalist, and before the development of large scale mining, smelting or shipbuilding industries. In these societies agriculture was the driving force in the economy and shifting cultivation was the most common type of agriculture practiced. By examining the relationships between social and economic change and agricultural change in these societies, insights can be gained on contemporary social and economic change and global environment change, and the place of shifting cultivation in those relationships.

As early as 1930 questions about relationships between the rise and fall of the Mayan civilization of the Yucatan Peninsula and shifting cultivation were raised and continue to be debated today. Archaeological evidence suggests the development of Mayan society and economy began around 250 AD. A mere 700 years later it reached its apogee, by which time the population may have reached 2,000,000 people. There followed a precipitous decline that left the great cities and ceremonial centres vacant and overgrown with jungle vegetation. The causes of this decline are uncertain; but warfare and the exhaustion of agricultural land are commonly cited (Meggers 1954; Dumond 1961; Turner 1974). More recent work suggests the Maya may have, in suitable places, developed irrigation systems and more intensive agricultural practices (Humphries 1993).

Similar paths appear to have been followed by Polynesian settlers in New Zealand and the Pacific Islands, who within 500 years of their arrival around 1100 AD turned substantial areas from forest into scrub and fern and in the process caused the elimination of numerous species of birds and animals (Kirch and Hunt 1997). In the restricted environments of the Pacific islands, including Fiji and Hawaii, early extensive erosion and change of vegetation is presumed to have been caused by shifting cultivation on slopes.

Soils washed from slopes were deposited in valley bottoms as a rich, swampy alluvium. These new environments were then exploited to develop intensive, irrigated fields. The change from shifting cultivation to intensive irrigated fields, occurred in association with a rapid growth in population and the development of elaborate and high stratified chiefdoms (Kirch 1984). In the larger, temperate latitude, islands of New Zealand the presumed course of events took a different path. There the stimulus for population growth was the hunting of large birds to extinction, during which time forests in drier areas were destroyed by burning, followed the development of intensive agriculture in favourable environments, based mainly on sweet potato (Ipomoea batatas) and a reliance on the gathering of two main wild plant species in less favourable environments. These changes, as in the smaller islands, were accompanied by population growth, the competition for the occupation of the best environments, complexity in social organisation, and endemic warfare (Anderson 1997).

The record of human induced changes in environments is longer in New Guinea than in most places. Agricultural activities probably beginning 5,000 to 9,000 years ago. However, the most spectacular changes, in both societies and environments, are believed to have occurred in the central highlands of the island within the last 1,000 years, in association with the introduction of a crop new to New Guinea, the sweet potato (Golson 1982a; 1982b). One of the most striking signals of the relatively recent intensification of agriculture is the sudden increase in sedimentation rates in small lakes.

The root question posed by these and the numerous other examples that could be cited of simple societies that have intensified their agricultural systems in association with increases in population and social complexity is not whether or how shifting cultivation was responsible for the extensive changes to landscapes and environments. Rather it is why simple societies of shifting cultivators in the tropical forest of Yucatan, or the highlands of New Guinea, begin to grow in numbers and to develop stratified and sometimes complex social hierarchies?

At first sight, the greatest stimulus to the intensification of a shifting cultivation system is a growth in population. If no other changes occur within the system, for each extra person to be fed from the system, a small extra amount of land must be cultivated. The total amount of land available is the land being presently cropped and all of the land in fallow. If the area occupied by the system is not expanded into previously unused land, then either the cropping period must be extended or the fallow period shortened.

At least two problems exist with the population growth hypothesis. First, population growth in most pre-industrial shifting cultivator societies has been shown to be very low over the long term. Second, no human societies are known where people work only to eat. People engage in social relations with each other and agricultural produce is used in the conduct of these relationships. These relationships are the focus of two attempts to understand the nexus between human societies and their environments, one an explanation of a particular situation and the other a general exploration of the problem.

In a study of the Duna in the Southern Highlands, a group in the process of moving from shifting cultivation into permanent field agriculture post sweet potato, Modjeska (1982) argued for the development of two "self amplifying feed back loops" of ecological and social causation. The trigger to the changes was very slow population growth and the slow expansion of agriculture to meet the demands of this growth. This set in motion the first feedback loop, the "use-value" loop. As more forest was cleared there was a decline in wild food resources and protein produced from hunting, which was substituted for by an increase in domestic pig raising. An increase in domestic pigs required a further expansion in agriculture. The greater protein available from the larger number of pigs increased human fertility and survival rates and resulted in faster population growth.

Increasing numbers of people also set in motion the second or "exchange-value" loop of social causalities. More people meant greater numbers of human interactions, including increased opportunities for conflict. There arose the need for a means to mediate these relationships. The Duna (and other highlanders in New Guinea) says Modjeska, substituted pigs for humans and began to exchange pigs to compensate for losses of humans in warfare, or in marriage and deaths from natural causes. Demand for pigs increased and the production of pigs increased to meet the demand, with all of the consequences observed in the use-value loop. This is what Brookfield (1972) called "social production". Increased pig production and the exchange of pigs to mediate relations between individuals and groups created opportunities for leadership and management of resources and some men gained authority over other men and all women. As groups became more complex, competition between men and between groups increased, increasing the opportunities for conflict.

The outcome of the operation of the two loops, one bringing about ecological change and the other social and economic change, is an expanding and intensifying agricultural system, the conversion of forest to grassland, a population growing at an increasing rate and expanding geographically and a society that is increasing in complexity and stratification. The second attempt to explain the relationships between simple agricultural societies and their environments is that of Ellen (1982, 252-270). Ellen does not attempt to separate use-values from social production. He argues that almost all of the materials required by humans to live (with perhaps the exception of air) are obtained through social relations of production and that these relations proliferate and are modified in numerous ways. The values that humans attribute to items produced from the environment arise out of cultural arrangements and not from the objects themselves, a restatement of Karl Sauer's dictum that "resources are cultural appraisals".

Humans frequently translate actual objects into culturally conceived forms, an example being the translation by the Duna of the pig into an item of compensation and redemption. As a result, two fundamental processes underlie the ecology of human social systems: First, the obtaining of materials from the environment and their alteration and circulation through social relations, and second, the giving of the material a value which will affect how important it is to obtain it, circulate it or alter it. Environmental pressures are thus mediated through social relations.

Transitions in ecological systems and in social systems do not proceed at the same rate. The rate of phylogenetic change is determined mainly by natural selection and partly by human interference and adaptation, such as for example, the domestication of a wild species. Humans however have the ability to learn and to communicate their knowledge to each other and across generations. If most social systems have the tendency to increase in complexity they will, sooner or later, come into conflict with, or into "contradiction" (Friedman 1979, 1982) with their environments. What happens around the point of "contradiction" will determine the extent of the environmental degradation that will occur. Of particular importance is the ability of the society to change, to invent or to innovate technologically and sociologically, in order to overcome the "contradiction" without incurring continuing environmental degradation, or social disintegration.

An economic study of what occurs at the points of conflict with specific reference to shifting cultivation is that of Ester Boserup (1965). Boserup argues that low intensity farming, extensive shifting cultivation for example, has lower labour costs than more intensive farming systems. This assertion remains controversial. She also argues that given a choice, a human group will always choose the technique which has the lowest absolute labour cost rather than the highest yield. But at the point of conflict, yields will have become unsatisfactory. Boserup argues, contra Malthus, that rather than population always overwhelming resources, that humans will invent a new agricultural technique or adopt an existing innovation that will boost yields and that is adapted to the new environmental conditions created by the degradation which has occurred already, even though they will pay for the increases in higher labour costs.

Examples of such changes are the adoption of new higher yielding crop, the exchanging of a digging stick for a hoe, or a hoe for a plough, or the development of irrigation systems. The controversy over Boserup's proposal is in part over whether intensive systems are more costly in labour terms, and whether humans will bring about change in their agricultural systems before environmental degradation forces them to. A number of very important things happen in the passage from simple to more complex societies and agricultural systems (Ellen 1982, 272-273). Demands for production on a local system by an external one may destabilise the local ability to regulate human environment relations.

Parts of the agricultural system may become more specialised, species diversity may be reduced or lost, wild plant and animal resources reduced or lost and ecosystems become more fragile. Improved communications result in a higher rate of innovation and hence a greater rate of change. Higher rates of change and increased differentiation in the society give rises to increased conflict. Increased differentiation also leads to larger numbers of individuals not producing anything, and more being produced by fewer, such that the system as a whole becomes less efficient. Greater organisation and specialisation results in greater complexity, technical division of labour and a greater codification of cultural responses with more extensive social control.

Shifting Cultivation in the Contemporary World and Global Environmental Change

The estimated rate of deforestation in South East Asia in 1990 was 34,000 sq km per year (FAO 1990, quoted in Potter 1993). In Indonesia alone it was estimated 13,100 sq km per year were being lost, 3,680 sq km per year from Sumatra and 3,770 sq km from Kalimantan, of which 1,440 sq km were due to the fires of 1982 to 1983. Since those estimates were made huge fires have ravaged Indonesian forests during the 1997 to 1998 El Nino associated drought. Efforts are being made in Indonesia to encourage shifting cultivators to alter the mix of activities and examine alternative cropping patterns, so that the slash and burn portion of their shifting cultivation is a smaller fraction of the time interval of the farming cycle.

For example by introducing jungle rubber farming instead of coffee, the farming cycle for growing rubber trees can extend up to 28 years versus about seven for coffee. The outcomes not only reflect less time spent in the slash and burn phase, but also allows a cover crop (rubber) that provides a forest habitat quality much higher than the coffee farm environment. Furthermore the coffee use often is abandoned entirely, yielding little of residual habitat and inviting a much earlier slash and burn element to recur.

Shifting cultivation was assessed by the FAO to be one a causes of deforestation while logging was not. The apparent discrimination against shifting cultivators caused a confrontation between FAO and environmental groups, who saw the FAO supporting commercial logging interests against the rights of indigenous people (Potter 1993, 108). Other independent studies of the problem note that despite lack of government control over forests and the dominance of a political elite in the logging industry, the causes of deforestation are more complex. The loggers have provided paid employment to former subsistence farmers. One of the outcomes of cash incomes has been rapid population growth among indigenous groups of former shifting cultivators that has placed pressure on their traditional long fallow farming systems. Many farmers have taken advantage of the improved road access to urban areas by planting cash crops, such as rubber or pepper as noted above.

Increased cash incomes often are spent on chain saws, which have enabled larger areas to be cleared for cultivation. Fallow periods have been reduced and cropping periods extended. Serious poverty elsewhere in the country has brought thousands of land hungry settlers into the cut over forests along the logging roads. The settlers practice what appears to be shifting cultivation but which is in fact a one-cycle slash and burn followed by continuous cropping, with no intention to long fallow. Clearing of trees and the permanent cultivation of fragile soils in a tropical environment with little attempt to replace lost nutrients may cause rapid degradation of the fragile soils.

The loss of forest in Indonesia, Thailand and the Philippines during the 1990s was preceded by major ecosystem disruptions in Vietnam, Laos and Cambodia in the 1970s and 1980s caused by warfare. Forests were sprayed with defoliants, thousands of rural forest dwelling people uproots from their homes and moved and roads driven into previously isolated areas. The loss of the tropical forests of South East Asia is the particular outcome of the general possible outcomes described by Ellen when small local ecological and social systems become part of larger system. When the previous relatively stable ecological relationships are destabilised, degradation can occur rapidly. Similar descriptions of the loss of forest and destruction of fragile ecosystems could be provided from the Amazon Basin, by large scale state sponsored colonisation forest land (Becker 1995, 61) or from the Central Africa where what endemic armed conflict is destabilising rural settlement and farming communities on a massive scale.

Comparison with other Ecological Phenomena

In the tropical developing world, shifting cultivation in its many diverse forms, remains a pervasive practice. Shifting cultivation was one of the very first forms of agriculture practiced by humans and its survival into the modern world suggests that it is a flexible and highly adaptive means of production. However, it is also a grossly misunderstood practice. Many casual observers cannot see past the clearing and burning of standing forest and do not perceive often ecologically stable cycles of cropping and fallowing. Nevertheless, shifting cultivation systems are particularly susceptible to rapid increases in population and to economic and social change in the larger world around them.

The blame for the destruction of forest resources is often laid on shifting cultivators. But the forces bringing about the rapid loss of tropical forests at the end of the 20th century are the same forces that led to the destruction of the forests of Europe, urbanisation, industrialisation and the application the latest technology to extract ever more resources from the environment in pursuit of political power by competing groups.

Studies of small, isolated and pre-capitalist groups and their relationships with their environments suggests that the roots of the contemporary problem lie deep in human behavioural patterns, for even in these simple societies, competition and conflict can be identified as the main force driving them into contradiction with their environments.

Alternative Practice in the Pre-Columbian Amazon Basin

Slash-and-char, as opposed to slash-and-burn, may create self-perpetuating soil fertility that supports sedentary agriculture, but the society so sustained may still be overturned.

Soil Contamination

Soil contamination is caused by the presence of man-made chemicals or other alteration in the natural soil environment.

This type of contamination typically arises from the rupture of underground storage tanks, application of pesticides, percolation of contaminated surface water to subsurface strata, oil and fuel dumping, leaching of wastes from landfills or direct discharge of industrial wastes to the soil. The most common chemicals involved are petroleum hydrocarbons, solvents, pesticides, lead and other heavy metals. This occurrence of this phenomenon is correlated with the degree of industrialisations and intensities of chemical usage.

The concern over soil contamination stems primarily from health risks, both of direct contact and from secondary contamination of water supplies. Mapping of contaminated soil sites and the resulting cleanup are time consuming and expensive tasks, requiring extensive amounts of geology, hydrology, chemistry and computer modelling skills.

It is in North America and Western Europe that the extent of contaminated land is most well known, with many of countries in these areas having a legal framework to identify and deal with this environmental problem; this however may well be just the tip of the iceberg with developing countries very likely to be the next generation of new soil contamination cases.

The immense and sustained growth of the People's Republic of China since the 1970s has exacted a price from the land in increased soil pollution. The State Environmental Protection Administration believes it to be a threat to the environment, to food safety and to sustainable agriculture. According to a scientific sampling, 150 million mi (100,000 square kilometres) of China's cultivated land have been polluted, with contaminated water being used to irrigate a further 32.5 million mi (21,670 square kilometres) and another 2 million mi (1,300 square kilometres) covered or destroyed by solid waste. In total, the area accounts for one-tenth of China's cultivatable land, and is mostly in economically developed areas. An estimated 12 million tonnes of grain are contaminated by heavy metals every year, causing direct losses of 20 billion yuan (US $2.57 billion).

The United States, while having some of the most widespread soil contamination, has actually been a leader in defining and implementing standards for cleanup. Other industrialised countries have a large number of contaminated sites, but lag the US in executing remediation. Developing countries may be leading in the next generation of new soil contamination cases.

Each year in the US, thousands of sites complete soil contamination cleanup, some by using microbes that "eat up" toxic chemicals in soil, many others by simple excavation and others by more expensive high-tech soil vapour extraction or air stripping. Efforts proceed worldwide to identify new sites of soil contamination.

Ecosystem Effects

Not unexpectedly, soil contaminants can have significant deleterious consequences for ecosystems. There are radical soil chemistry changes which can arise from the presence of many hazardous chemicals even at low concentration of the contaminant species. These changes can manifest in the alteration of metabolism of endemic microorganisms and arthropods resident in a given soil environment. The result can be virtual eradication of some of the primary food chain, which in turn have major consequences for predator or consumer species. Even if the chemical effect on lower life forms is small, the lower pyramid levels of the food chain may ingest alien chemicals, which normally become more concentrated for each consuming rung of the food chain. Many of these effects are now well known, such as the concentration of persistent DDT materials for avian consumers, leading to weakening of egg shells, increased chick mortality and potentially species extinction.

Effects occur to agricultural lands which have certain types of soil contamination. Contaminants typically alter plant metabolism, most commonly to reduce crop yields. This has a secondary effect upon soil conservation, since the languishing crops cannot shield the earth's soil mantle from erosion phenomena. Some of these chemical contaminants have long half-lives and in other cases derivative chemicals are formed from decay of primary soil contaminants.

Regulatory Framework

United States of America

Until about 1970 there was little widespread awareness of the worldwide scope of soil contamination or its health risks. In fact, areas of concern were often viewed as unusual or isolated incidents. Since then, the US has established guidelines for handling hazardous waste and the cleanup of soil pollution. In 1980 the US Superfund/CERCLA established strict rules on legal liability for soil contamination. Not only did CERCLA stimulate identification and cleanup of thousands of sites, but it raised awareness of property buyers and sellers to make soil pollution a focal issue of land use and management practices.

While estimates of remaining soil cleanup in the US may exceed 200,000 sites, in other industrialised countries there is a lag of identification and cleanup functions. Even though their use of chemicals is lower than industrialised countries, often their controls and regulatory framework is quite weak. For example, some persistent pesticides that have been banned in the US are in widespread uncontrolled use in developing countries. It is worth noting that the cost of cleaning up a soil contaminated site can range from as little as about $10,000 for a small spill, which can be simply excavated, to millions of dollars for a widespread event, especially for a chemical that is very mobile such as perchloroethylene.

United Kingdom

Generic guidance commonly used in the UK are the Soil Guideline Values published by DEFRA and the Environment Agency. These are screening values that demonstrate the minimal acceptable level of a substance. Above this there can be no assurances in terms of significant risk of harm to human health. These have been derived using the Contaminated Land Exposure Assessment Model (CLEA UK). Certain input parameters such as Health Criteria Values, age and land use are fed into CLEA UK to obtain a probablistic output.

Guidance by the Inter Departmental Committee for the Redevelopment of Contaminated Land (ICRCL) has been formally withdrawn by the Department for Environment, Food and Rural Affairs (DEFRA), for use as a prescriptive document to determine the potential need for remediation or further assessment. Therefore, no further reference is made to these former guideline values.

Other generic guidance that exists (to put the concentration of a particular contaminant in context), includes the United States EPA Region 9 Preliminary Remediation Goals (US PRGs), the US EPA Region 3 Risk Based Concentrations (US EPA RBCs) and National Environment Protection Council of Australia Guideline on Investigation Levels in Soil and Groundwater.

However international guidance should only be used in the UK with clear justification. This is because foreign standards are usually particular to that country due to drivers such as political policy, geology, flood regime and epidemiology. It is generally accepted by UK regulators that only robust scientific methods that relate to the UK should be used.

The CLEA model published by DEFRA and the Environment Agency (EA) in March 2002 sets a framework for the appropriate assessment of risks to human health from contaminated land, as required by Part IIA of the Environmental Protection Act 1990. As part of this framework, generic Soil Guideline Values (SGVs) have currently been derived for ten contaminants to be used as "intervention values". These values should not be considered as remedial targets but values above which further detailed assessment should be considered.

Three sets of CLEA SGVs have been produced for three different land uses, namely:

- residential (with and without plant uptake);
- allotments;
- commercial/industrial.

It is intended that the SGVs replace the former ICRCL values. It should be noted that the CLEA SGVs relate to assessing chronic (long term) risks to human health and do not apply to the protection of ground workers during construction, or other potential receptors such as groundwater, buildings, plants or other ecosystems. The CLEA SGVs are not directly applicable to a site completely covered in hardstanding, as there is no direct exposure route to contaminated soils.

To date, the first ten of fifty-five contaminant SGVs have been published, for the following: arsenic, cadmium, chromium, lead, inorganic mercury, nickel, selenium ethyl benzene, phenol and toluene. Draft SGVs for benzene, naphthalene and xylene have been produced but their publication is on hold. Toxicological data (Tox) has been published for each of these contaminants as well as for benzo[a]pyrene, benzene, dioxins, furans and dioxin-like PCBs, naphthalene, vinyl chloride, 1,1,2,2 tetrachloroethane and 1,1,1,2 tetrachloroethane, 1,1,1 trichloroethane, tetrachloroethene, carbon tetrachloride, 1,2-dichloroethane, trichloroethene and xylene. The SGVs for ethyl benzene, phenol and toluene are dependent on the soil organic matter (SOM) content (which can be calculated from the total organic carbon (TOC) content). As an initial screen the SGVs for 1% SOM are considered to be appropriate.

Cleanup Options

Cleanup or remediation is analysed by environmental scientists who utilise field measurement of soil chemicals and also apply computer models for analysing transport and fate of soil chemicals. Thousands of soil contamination cases are currently in active cleanup across the US as of 2006. There are several principal strategies for remediation:

- Excavate soil and take it to a disposal site away from ready pathways for human or sensitive ecosystem contact. This technique also applies to dredging of bay muds containing toxins.
- Aeration of soils at the contaminated site (with attendant risk of creating air pollution)

- Thermal remediation by introduction of heat to raise subsurface temperatures sufficiently high to volatise chemical contaminants out of the soil for vapour extraction. Technologies include ISTD, electrical resistance heating (ERH), and ET-DSP[tm].
- Bioremediation, involving microbial digestion of certain organic chemicals. Techniques used in bioremediation include landfarming, biostimulation and bioaugmentating soil biota with commercially available microflora.
- Extraction of groundwater or soil vapour with an active electromechanical system, with subsequent stripping of the contaminants from the extract.
- Containment of the soil contaminants (such as by capping or paving over in place).

Soil Life

Soil life or soil biota is a collective term for all the organisms living within the soil.

Overview

In a balanced soil, plants grow in an active and vibrant environment. The mineral content of the soil and its physical structure are important for their well-being, but it is the life in the earth that powers its cycles and provides its fertility. Without the activities of soil organisms, organic materials would accumulate and litter the soil surface, and there would be no food for plants. The soil biota includes:

- Megafauna: size range 20 mm upwards, e.g. moles, rabbits, and rodents.
- Macrofauna: size range 2-20 mm, e.g. woodlice, earthworms, beetles, centipedes, slugs, snails, ants, and harvestmen.
- Mesofauna: size range 100 micrometre-2 mm, e.g. tardigrades, mites and springtails.
- Microfauna and Microflora: size range 1-100 micrometres, e.g. yeasts, bacteria (commonly actinobacteria), fungi, protozoa, roundworms, and rotifers.

Of these, bacteria and fungi play key roles in maintaining a healthy soil. They act as decomposers that break down organic materials to produce detritus and other breakdown products. Soil detritivores, like earthworms, ingest detritus and decompose it. Saprotrophs, well represented by fungi and bacteria, extract soluble nutrients from delitro.

Bacteria

Bacteria are single-celled organisms, and are the most numerous denizens of the soil, with populations ranging from 100 million to 3 billion in a gram. They are capable of very rapid reproduction by binary fission (dividing into two) in favourable conditions. One bacterium is capable of producing 16 million more in just 24 hours. Most soil bacteria live in close proximity to plant roots and are often referred to as rhizobacteria. Bacteria live in soil water, including the film of moisture surrounding soil particles, and some are able to swim by means of flagella. The majority of the beneficial soil-dwelling bacteria need oxygen (and are thus termed aerobic bacteria), whilst those that do not require air are referred to as anaerobic, and tend to cause putrefaction of dead organic matter. Aerobic bacteria are most active in a soil that is moist (but not saturated, as this will deprive aerobic bacteria of the air that they require), and neutral soil pH, and where there is plenty of food (carbohydrates and micronutrients from organic matter) available. Hostile conditions will not completely kill bacteria; rather, the bacteria will stop growing and get into a dormant stage, and those individuals with pro-adaptive mutations may compete better in the new conditions. Gram positive bacteria produce spores in order to wait for more favourable circumstances, and Gram negative bacteria gets into a "nonculturable" stage.

From the organic gardener's point of view, the important roles that bacteria play are:

Nitrification

Nitrification is a vital part of the nitrogen cycle wherein certain bacteria (which manufacture their own carbohydrate supply without using the process of photosynthesis) are able to transform nitrogen in the form of ammonium, which is produced by the decomposition of proteins, into nitrates, which are available to growing plants, and once again converted to proteins.

Nitrogen Fixation

In another part of the cycle, the process of nitrogen fixation constantly puts additional nitrogen into biological circulation. This is carried out by free-living nitrogen-fixing bacteria in the soil or water such as *Azotobacter*, or by those which live in close symbiosis with leguminous plants, such as rhizobia. These bacteria form colonies in nodules they create on the roots of peas, beans, and related species. These are able to convert nitrogen from the atmosphere into nitrogen-containing organic substances.

Denitrification

While nitrogen fixation converts nitrogen from the atmosphere into organic compounds, a series of processes called denitrification returns an approximately equal amount of nitrogen to the atmosphere. Denitrifying bacteria tend to be anaerobes, or facultatively anaerobes (can alter between the oxygen dependent and oxygen independent types of metabolisms), including *Achromobacter* and *Pseudomonas*. The putrefaction process caused by oxygen-free conditions converts nitrates and nitrites in soil into nitrogen gas or into gaseous compounds such as nitrous oxide or nitric oxide. In excess, denitrification can lead to overall losses of available soil nitrogen and subsequent loss of soil fertility. However, fixed nitrogen may circulate many times between organisms and the soil before denitrification returns it to the atmosphere.

Actinobacteria

Actinobacteria are critical in the decomposition of organic matter and in humus formation, and their presence is responsible for the sweet "earthy" aroma which is associated with a good healthy soil. They require plenty of air and a pH between 6.0 and 7.5, but are more tolerant of dry conditions than most other bacteria and fungi.

Fungi

A gram of garden soil can contain around one million fungi, such as yeasts and moulds. Fungi have no chlorophyll, and are not able to photosynthesise; besides, they can't use atmospheric carbon dioxide as a source of carbon, therefore they are chemo-heterotrophic, meaning that, like animals, they require a chemical source of energy rather than being able to use light as an energy source, as well as organic substrates to get carbon for growth and development.

Many fungi are parasitic, often causing disease to their living host plant, although some have beneficial relationships with living plants. In terms of soil and humus creation, the most important fungi tend to be saprotrophic, that is, they live on dead or decaying organic matter, thus breaking it down and converting it to forms which are available to the higher plants. A succession of fungi species will colonise the dead matter, beginning with those that use sugars and starches, which are succeeded by those that are able to break down cellulose and lignins.

Fungi spread underground by sending long thin threads known as mycelium throughout the soil; these threads can be observed throughout many soils and compost heaps. From the mycelia the fungi is able to throw up its fruiting bodies, the visible part above the soil (e.g., mushrooms, toadstools and puffballs) which may contain millions of spores. When the fruiting body bursts, these spores are dispersed through the air to settle in fresh environments, and are able to lie dormant for up to years until the right conditions for their activation arise or the right food is made available.

Mycorrhizae

Those fungi that are able to live symbiotically with living plants, creating a relationship that is beneficial to both, are known as Mycorrhizae (from *myco* meaning fungal and *rhiza* meaning root). Plant root hairs are invaded by the mycelia of the mycorrhiza, which lives partly in the soil and partly in the root, and may either cover the length of the root hair as a sheath or be concentrated around its tip. The mycorrhiza obtains the carbohydrates that it requires from the root, in return providing the plant with nutrients including nitrogen and moisture. Later the plant roots will also absorb the mycelium into its own tissues.

Beneficial mycorrhizal associations are to be found in many of our edible and flowering crops. Shewell Cooper suggests that these include at least 80% of the *brassica* and *solanum* families (including tomatoes and potatoes), as well as the majority of tree species, especially in forest and woodlands. Here the mycorrhizae create a fine underground mesh which extends greatly beyond the limits of the tree's roots, thus greatly increasing their feeding range and actually causing neighbouring trees to become physically interconnected. The benefits of mycorrhizal relations to their plant partners are not limited to nutrients, but can be essential for plant reproduction: in situations where little light is able to reach the forest floor, such as the North American pine forests, a young seedling cannot obtain sufficient light to photosynthesise for itself and will not grow properly in a sterile soil. But if the ground is underlain by a mycorrhizal mat then the developing seedling will throw down roots that can link with the fungal threads and through them obtain the nutrients it needs, often indirectly obtained from its parents or neighbouring trees. Shel organics giving organic manure in the trade name of SOIL RICH available in Guntur with the above said properties

David Attenborough points out the plant/fungi/animal relationship that creates a "Three way harmonious trio" to be found in forest ecosystems wherein the plant/fungi symbiosis is enhanced by animals such as the wild boar, deer, mice or flying squirrel which feed upon the fungi's fruiting bodies, including truffles, and cause their further spread (*Private Life Of Plants*, 1995). A greater understanding of the complex relationships which pervade natural systems is one of the major justifications of the organic gardener, in refraining from the use of artificial chemicals and the damage these might cause.

17

Irrigation Water Management

Concept and Perception

All field crops need soil, water, air and light (sunshine) to grow. The soil gives stability to the plants; it also stores the water and nutrients which the plants can take up through their roots. The sunlight provides the energy which is necessary for plant growth. The air allows the plants to "breath".

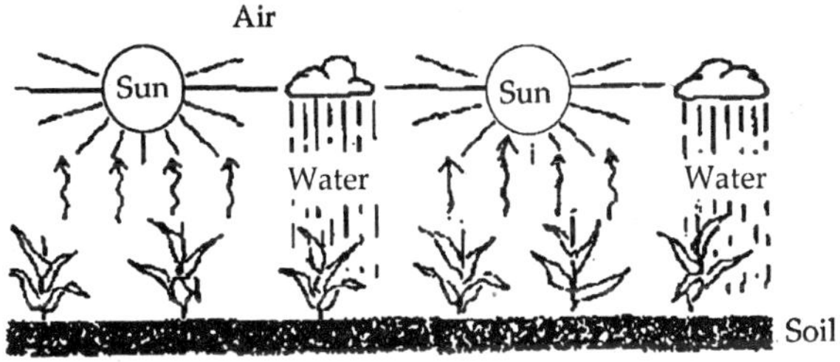

Figure: Plants need Soil, Water, Air and Sunlight.

Without water crops cannot grow. Too much water is not good for many crops either. Apart from paddy rice, there are only very few crops which like to grow "with their feet in the water". The most well-known source of water for plant growth is rain water. There are two important questions which come to mind: What to do if there is too much rain water? What to do if there is too little rain water?

If there is too much rain, the soil will be full of water and there will not be enough air. Excess water must be removed.

The removal of excess water - either from the ground surface or from the root zone - is called drainage.

If there is too little rain, water must be supplied from other sources; irrigation is needed. The amount of irrigation water which is needed depends not only on the amount of water already available from rainfall, but also on the total amount of water needed by the various crops.

With respect to the need for irrigation water, a distinction can be made among three climatic situations:

1. *Humid Climates:* More than 1200 mm of rain per year. The amount of rainfall is sufficient to cover the water needs of the various crops. Excess water may cause problems for plant growth and thus drainage is required.
2. *Sub-humid and Semi-arid Climates:* Between 400 and 1200 mm of rain per year. The amount of rainfall is important but often not sufficient to cover the water needs of the crops. Crop production in the dry season is only possible with irrigation, while crop production in the rainy season may be possible but unreliable: yields will be less than optimal.
3. *Semi-arid, Arid and Desert Climates:* Less than 400 mm of rain per year. Reliable crop production based on rainfall is not possible; irrigation is thus essential.

The two major factors which determine the amount of irrigation water which is needed are:

- the total water need of the various crops;
- the amount of rain water which is available to the crops.

In other words, the irrigation water need is the difference between the total water need of the crops and the amount of rainfall which is available to the crops.

Crop Water Need − Available Rain Water = Irrigation Water Need

Figure: Irrigation Water Need

In many countries it is already well known what the crop water needs and irrigation water needs are of the most commonly grown crops. Such data can usually be obtained from the Extension Service, the Irrigation Department or Ministry of Agriculture. It is then not necessary to determine the crop and irrigation water need. However, there may be situations where it is not possible to obtain these data and it would thus be necessary to determine them on the spot.

Part I of this manual allows the reader to make a rough estimation of the crop and irrigation water needs - without using any complicated calculations.

Part II of this manual allows the reader actually to calculate - be it in a fairly simple manner - the crop and irrigation water needs. These calculations obviously lead to a greater accuracy, but they also require more time and background knowledge.

Crop Water Needs

The plant roots suck or extract water from the soil to live and grow. The main part of this water does not remain in the plant, but escapes to the atmosphere as vapour through the plant's leaves and stem. This process is called transpiration. Transpiration happens mainly during the day time.

Water from an open water surface escapes as vapour to the atmosphere during the day. The same happens to water on the soil surface and to water on the leaves and stem of a plant. This process is called evaporation.

The water need of a crop thus consists of transpiration plus evaporation. Therefore, the crop water need is also called "evapotranspiration".

The water need of a crop is usually expressed in mm/day, mm/month or mm/season.

Suppose the water need of a certain crop in a very hot, dry climate is 10 mm/day. This means that each day the crop needs a water layer of 10 mm over the whole area on which the crop is grown. It does not mean that this 10 mm has to indeed be supplied by rain or irrigation every day.

It is, of course, still possible to supply, for example, 50 mm of irrigation water every 5 days. The irrigation water will then be stored in the root zone and gradually be used by the plants: every day 10 mm.

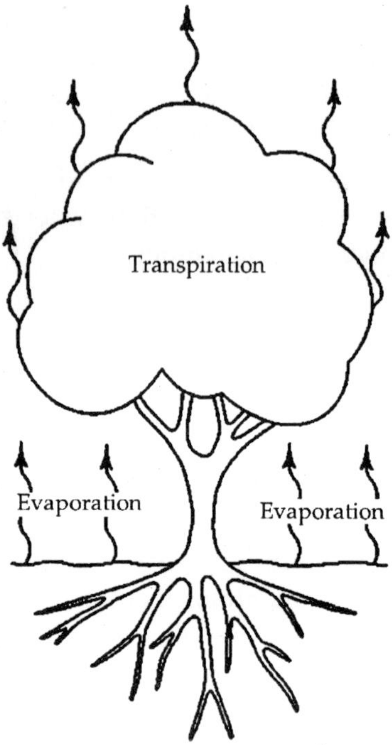

Figure: Evapotranspiration

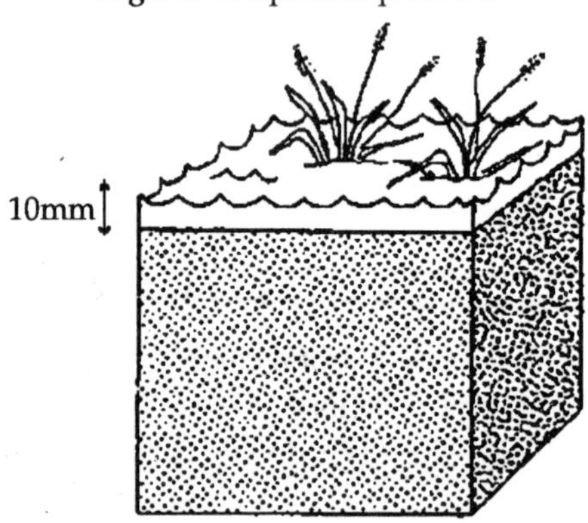

Figure: A Crop Water Need of 10 mm/day

The crop water need mainly depends on:

- *The Climate:* For example, in a sunny and hot climate crops need more water per day than in a cloudy and cool climate.
- *The Crop Type:* Crops like rice or sugarcane need more water than crops like beans and wheat.
- *The Growth Stage:* Grown crops need more water than crops that have just been planted.

Influence of the Climate on Crop Water Needs

A certain crop grown in a sunny and hot climate needs per day more water than the same crop grown in a cloudy and cooler climate. There are, however – apart from sunshine and temperature – other climatic factors which influence the crop water need. These factors are the humidity and the windspeed. When it is dry, the crop water needs are higher than when it is humid. In windy climates the crops will use more water than in calm climates.

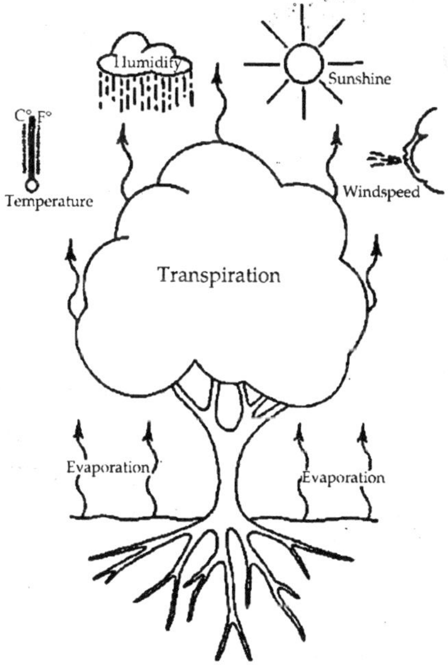

Figure: Major Climatic Factors Influencing Crop Water Needs

The effect of these four climatic factors on the water need of the crop is shown in following table.

Table: Effect of Major Climatic Factors on Crop Water Needs

Climatic Factor	*Crop water need*	
	High	*Low*
Sunshine	sunny (no clouds)	cloudy (no sun)
Temperature	hot	cool
Humidity	low (dry)	high (humid)
Windspeed	windy	little wind

From the above it is clear that one crop grown in different climatic zones will have different water needs. For example, a certain maize variety grown in a cool climate will need less water per day than the same maize variety grown in a hotter climate.

It is therefore useful to take a certain standard crop or reference crop and determine how much water this crop needs per day in the various climatic regions. As a standard crop or reference crop grass has been chosen.

Following table indicates the average daily water needs of this reference grass crop. The daily water needs of the grass depend on the climatic zone (rainfall regime) and daily temperatures.

Table: Average Daily Water Need of Standard Grass During Irrigation Season

Climatic zone	*Mean daily temperature*		
	low (less than 15°C)	*medium (15-25°C)*	*high (more than 25°C)*
Desert/arid	4-6	7-8	9-10
Semi arid	4-5	6-7	8-9
Sub-humid	3-4	5-6	7-8
Humid	1-2	3-4	5-6

For example, the standard grass crop grown in a semi-arid climate with a mean temperature of 20°C needs approximately 6.5 mm of water per day. The same grass crop grown in a sub-humid climate with a mean temperature of 30°C needs some 7.5 mm of water per day.

This daily water need of the standard grass crop is also called "reference crop evapotranspiration".

Influence of the Crop Type on the Crop Water Needs

The influence of the crop type on the crop water need is important in two ways:

1. The crop type has an influence on the daily water needs of a fully grown crop; i.e. the peak daily water needs: a fully developed maize crop will need more water per day than a fully developed crop of onions.
2. The crop type has an influence on the duration of the total growing season of the crop. There are short duration crops, e.g. peas, with a duration of the total growing season of 90-100 days and longer duration crops, e.g. melons, with a duration of the total growing season of 120-160 days. And then there are, of course, the perennial crops that are in the field for many years, such as fruit trees.

While, for example, the daily water need of melons may be less than the daily water need of peas, the seasonal water need of melons will be higher than that of beans because the duration of the total growing season of melons is much longer.

Influence of Crop Type on the Daily Crop Water Needs

Earlier, it has been indicated how the daily water need of standard grass can be estimated. In this section, it will be explained how the daily water needs of other crops can be estimated using as a basis the daily water need of the standard grass.

It will be easy to understand that a fully grown maize crop — with its large leaf area — will use more water per day than, for example, a fully grown crop of radishes or onions; that is when the two crops are grown in the same area.

When determining the influence of the crop type on the daily crop water needs, reference is always made to a fully grown crop; the plants have reached their maximum height; they optimally cover the ground; they possibly have started flowering or started grain setting. When the crops are fully grown their water need is the highest. It is the so-called "peak period" of their water needs.

For the various field crops it is possible to determine how much water they need compared to the standard grass. A number of crops need less water than grass, a number of crops need more water than grass and a number of crops need more or less the same amount of water as grass.

Following table indicates five groups of crops. The crops in column 1 need 30 per cent less water than grass in their peak period. The crops in column 2 need 10 per cent less water than grass. The crops in column 3 need the same amount of water as grass. The crops in columns 4 and 5 need respectively 10 and 20 per cent more water than grass in their peak period.

Table: Crop Water Needs in Peak Period of Various Field Crops as Compared to Standard Grass

Column 1	*Column 2*	*Column 3*	*Column 4*	*Column 5*
-30%	-10%	same as standard grass	+ 10%	+20%
citrus	cucumber	carrots	barley	paddy rice
olives	radishes	crucifers (cabbage, cauliflower, broccoli, etc.)	beans	sugarcane
grapes	squash	lettuce	maize	banana
		melons	flax	nuts and fruit trees with cover crop
		onions	small grains	
		peanuts	cotton	
		peppers	tomato	
		spinach	eggplant	

Contd...

Column 1	*Column 2*	*Column 3*	*Column 4*	*Column 5*
		tea	lentils	
		grass	millet	
		cacao	oats	
		coffee	peas	
		clean cultivated nuts and fruit trees, e.g. apples	potatoes	
			safflower	
			sorghum	
			soya beans	
			sugarbeet	
			sunflower	
			tobacco	
			wheat	

For example, suppose in a certain area the standard grass crop needs 5.5 mm of water per day.

Then, in that same area, maize will need 10 per cent more water. Ten per cent of 5.5 mm = 10/100 × 5.5 = 0.55 mm. Thus maize would need 5.5 + 0.55 = 6.05 or rounded 6.1 mm of water per day.

Influence of Crop Type on the Seasonal Crop Water Needs

The crop type not only has an influence on the daily water need of a fully grown crop, i.e. the daily peak water need, but the crop type also has an influence on the duration of the total growing season of the crop, and thus on the seasonal water need.

Data on the duration of the total growing season of the various crops grown in an area can best be obtained locally. These data may be obtained from, for example, the seed supplier, the Extension Service, the Irrigation Department or Ministry of Agriculture.

The duration of the total growing season has an enormous influence on the seasonal crop water need. There are, for example, many rice varieties, some with a short growing cycle (e.g. 90 days) and others with a long growing cycle (e.g. 150 days). This has a strong influence on the seasonal rice water needs: a rice crop which is in the field for 150 days will need in total much more water than a rice crop which is only in the field for 90 days. Of course, for the two rice crops the daily peak water need may still be the same, but the 150 day crop will need this daily amount for a longer period. The time of the year during which crops are grown is also very important. A certain crop variety grown during the cooler months will need substantially less water than the same crop variety grown during the hotter months.

Following table gives some Indicative values or approximate values for the duration of the total growing season for the various field crops. It should, however, be noted that these values are only rough approximations and it is much better to obtain the values locally.

Table: Indicative Values of the Total Growing Period

Crop	*Total growing period (days)*	*Crop*	*Total growing period (days)*
Alfalfa	100-365	Millet	105-140
Banana	300-365	Onion green	70-95
Barley/Oats/ Wheat	120-150	Onion dry	150-210
Bean green	75-90	Peanut/Groundnut	130-140
Bean dry	95-110	Pea	90-100
Cabbage	120-140	Pepper	120-210
Carrot	100-150	Potato	105-145
Citrus	240-365	Radish	35-45
Cotton	180-195	Rice	90-150
Cucumber	105-130	Sorghum	120-130
Eggplant	130-140	Soya bean	135-150
Flax	150-220	Spinach	60-100

Contd...

Crop	*Total growing period (days)*	*Crop*	*Total growing period (days)*
Grain/small	150-165.	Squash	95-120
Lentil	150-170	Sugarbeet	160-230
Lettuce	75-140	Sugarcane	270-365
Maize sweet	80-110	Sunflower	125-130
Maize grain	125-180	Tobacco	130-160
Melon	120-160	Tomato	135-180

As can be seen from above table there is a large variation of values not only between crops, but also within one crop type. In general it can be assumed that the growing period for a certain crop is longer when the climate is cool and shorter when the climate is warm.

Influence of the Growth Stage of the Crop on Crop Water Needs

A fully grown maize crop will need more water than a maize crop which has just been planted.

The crop water need or crop evapotranspiration consists of transpiration by the plant and evaporation from the soil and plant surface. When the plants are very small the evaporation will be more important than the transpiration. When the plants are fully grown the transpiration is more important than the evaporation.

Following figure shows in a schematic way the various development or growth stages of a crop.

At planting and during the initial stage, the evaporation is more important than the transpiration and the evapotranspiration or crop water need during the initial stage is estimated at 50 per cent of the crop water need during the mid - season stage, when the crop is fully developed.

During the so-called crop development stage the crop water need gradually Increases from 50 per cent of the maximum crop water need to the maximum crop water need.

The maximum crop water need is reached at the end of the crop development stage which is the beginning of the mid-season stage.

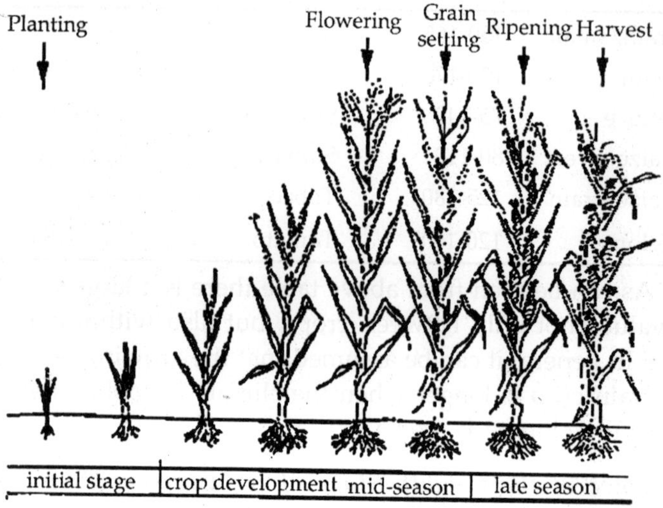

Figure: Growth Stages of a Crop

With respect to the late season stage, which is the period during which the crop ripens and is harvested, a distinction can be made between two groups of crops:

> Fresh harvested crops: such as lettuce, cabbage, etc. With these crops the crop water need remains the same during the late season stage as it was during the mid-season stage. The crops are harvested fresh and thus need water up to the last moment.

Dry harvested crops: such as cotton, maize (for grain production), sunflower, etc. During the late season stage these crops are allowed to dry out and sometimes even die. Thus their water needs during the late season stage are minimal. If the crop is indeed allowed to die, the water needs are only some 25 per cent of the crop water need during the mid-season or peak period. Of course, no irrigation is given to these crops during the late season stage.

Determination of Crop Water Needs

To calculate the water needs for the various months during which the crop is grown is fairly complicated and explained in Part II of this manual.

As stated before, it is often possible to obtain data on crop water needs locally and it is thus not necessary to calculate them. However, to give the reader some idea on values of seasonal water needs for the most important field crops.

Table: Approximate Values of Seasonal Crop Water Needs

Crop	*Crop water need (mm/total growing period)*
Alfalfa	800-1600
Banana	1200-2200
Barley/Oats/Wheat	450-650
Bean	300-500
Cabbage	350-500
Citrus	900-1200
Cotton	700-1300
Maize	500-800
Melon	400-600
Onion	350-550
Peanut	500-700
Pea	350-500
Pepper	600-900
Potato	500-700
Rice (paddy)	450-700
Sorghum/Millet	450-650
Soya bean	450-700
Sugarbeet	550-750
Sugarcane	1500-2500
Sunflower	600-1000
Tomato	400-800

Effective Rainfall

Apart from soil, air and sunlight, crops need water to grow. This water can be supplied to the crops by rainfall (also called precipitation), by irrigation or by a combination of rainfall and irrigation.

If the rainfall is sufficient to cover the water needs of the crops, irrigation is not required.

If there is no rainfall, all the water that the crops need has to be supplied by irrigation.

If there is some rainfall, but not enough to cover the water needs of the crops, irrigation water has to supplement the rain water in such a way that the rain water and the irrigation water together cover the water needs of the crop. This is often called supplemental irrigation: the irrigation water supplements or adds to the rain water.

As earlier explained, not all rain water which falls on the soil surface can indeed by used by the plants.

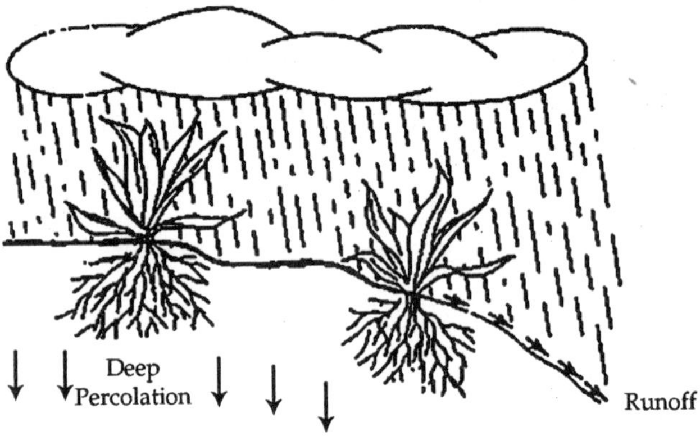

Figure: Part of the Rain Water is Lost through Deep Percolation and Run-off

Part of the rain water percolates below the root zone of the plants and part of the rain water flows away over the soil surface as run-off. This deep percolation water and run-off water cannot be used by the plants. In other words, part of the rainfall is not effective. The remaining part is stored in the root zone and can be used by the plants. This remaining part is the so-called effective rainfall. The factors which influence which part is effective and which part is not effective include the climate, the soil texture, the soil structure and the depth of the root zone.

If the rainfall is high, a relatively large part of the water is lost through deep percolation and run-off.

Deep Percolation: If the soil is still wet when the next rain occurs, the soil will simply not be able to store more water, and the rain water will thus percolate below the root zone and eventually reach the groundwater. Heavy rainfall may cause the groundwater table to rise temporarily.

Run-off: Especially in sloping areas, heavy rainfall will result in a large percentage of the rainwater being lost by surface run-off.

Another factor which needs to be taken into account when estimating the effective rainfall is the variation of the rainfall over the years. Especially in low rainfall climates, the little rain that falls is often unreliable; one year may be relatively dry and another year may be relatively wet.

In many countries, formulae have been developed locally to determine the effective precipitation. Such formulae take into account factors like rainfall reliability, topography, prevailing soil type, etc. If such formulae or other local data are available, they should be used.

If such data are not available. Following table could be used to obtain a rough estimate of the effective rainfall.

Table: Rainfall or Precipitation (P) and Effective Rainfall or Effective Precipitation (Pe) in MM/Month

P (mm/month)	*Pe (mm/month)*	*P (mm/month)*	*Pe (mm/month)*
0	0	130	79
10	0	140	87
20	2	150	95
30	8	160	103
40	14	170	111

Contd...

P (mm/month)	Pe (mm/month)	P (mm/month)	Pe (mm/month)
50	20	180	119
60	26	190	127
70	32	200	135
80	39	210	143
90	47	220	151
100	55	230	159
110	63	240	167
120	71	250	175

For example, estimate the effective rainfall in mm/month if the rainfall is 60 mm/month. From above table it can be seen that the effective rainfall is 26 mm/month. This means that out of 60 mm/month, some 26 mm can be used by the plants; and it is estimated that the remaining (60 - 26 =) 34 mm is lost through deep percolation and run-off.

Irrigation Water Needs

The irrigation water need of a certain crop is the difference between the crop water need and that part of the rainfall which can be used by the crop (the effective rainfall).

For each of the crops grown on an irrigation scheme the crop water need is determined, usually on a monthly basis; the crop water need is expressed in mm water layer per time unit, in this case mm/month.

The effective precipitation is estimated on a monthly basis, using measured rainfall data and above table (or local information, if available).

For all crops and for each month of the growing season, the irrigation water need is calculated by subtracting the effective rainfall from the crop water need.

Example: Suppose a tomato crop grown in a certain area has a total growing season of 150 days and the following monthly crop water needs:

	Feb	*Mar*	*Apr*	*May*	*June*	*Total*
Crop water need (mm/month)	69	123	180	234	180	786

This means that in February the tomatoes need 69 mm of water, in March 123 mm of water, etc. The water need of tomatoes over the total growing season (February-June: 150 days) is 786 mm.

Suppose the following rainfall data for the area where the tomatoes are grown have been obtained from the Meteorological Service or Ministry of Agriculture.

	Feb	*Mar*	*Apr*	*May*	*June*	*Total*
Rainfall: P (mm/month)	20	38	40	80	16	194

This means that the average rainfall for February is 20 mm, for March 38 mm, etc. The rainfall over the total growing season of tomatoes (February-June: 150 days) is 194 mm.

Only part of this rainfall is effective, and the effective rainfall is estimated earlier.

	Feb	*Mar*	*Apr*	*May*	*June*	*Total*
Rainfall: P (mm/month)	20	38	40	80	16	194
Effective rainfall: Pe (mm/month)	2	13	14	39	0	68

This means that the effective rainfall during February is only 2 mm, during April 13 mm, etc. The effective rainfall during the total growing season of tomatoes (February-June: 150 days) is 68 mm.

Now the Irrigation water need for the tomatoes can be calculated on a monthly basis, as follows:

	Feb	*Mar*	*Apr*	*May*	*June*	*Total*
Crop water need (mm/month)	69	123	180	234	180	786
Effective rainfall: Pe (mm/month)	2	13	14	39	0	68
Irrigation water need (mm/month)	67	110	166	195	180	718

Looking at the example for the month March, it can be seen that tomatoes need 123 mm during March. Of this 123 mm, 13 mm is supplied by the rainfall. The remaining (123 -13 =) 110 mm have to be supplied by irrigation. The total water need of tomatoes over the entire growing season is 786 mm of which 68 mm is supplied by rainfall. The remaining quantity (786 - 68 = 718 mm) has to be supplied by irrigation.

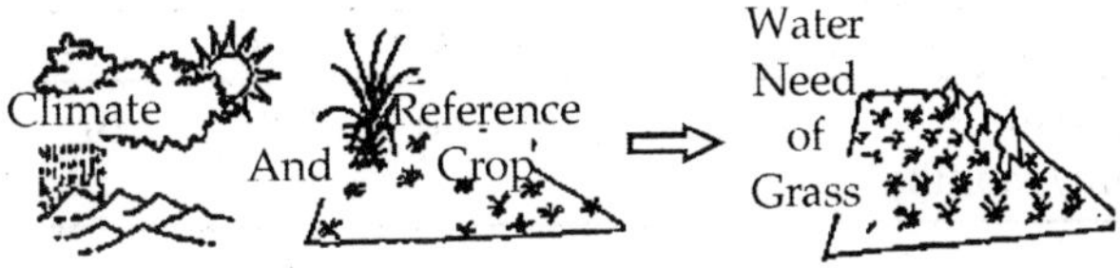

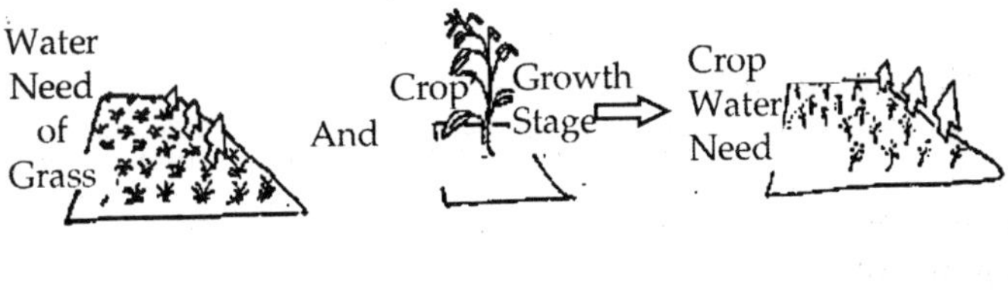

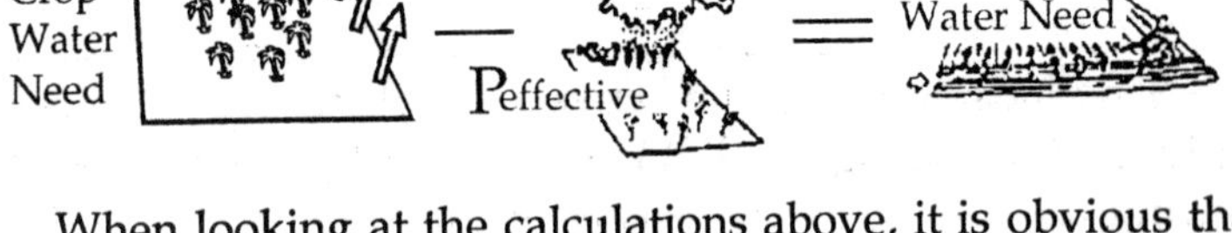

When looking at the calculations above, it is obvious that the month May is the month of peak irrigation water need (195

mm Irrigation water in May). If the tomatoes would be the only crop grown on the Irrigation scheme, the canals would have to be designed in such a way that they allow a flow large enough to supply indeed a net water layer of 195 mm to the whole area covered by tomatoes during the month May. In other words, for designing an irrigation scheme, the month of peak water supply is the critical month.

How to go about determining the scheme irrigation water supply and how to divide the irrigation water between the various crops and fields will be dealt with in Volume 4 of the Irrigation Water Management Training Manuals.

Major Climatic Zones

The major climatic factor which influences crop growth is rainfall. In areas with little rainfall the natural vegetation is limited, while in areas with high annual rainfall dense tropical rain forests are found.

Based on the annual rainfall, a distinction can be made between 6 major climatic zones as is shown in following table.

Table: Major Climatic Zones

Climatic Zone	*Annual Rainfall (mm)*	*Wet period (months)*	*Vegetation*
Desert	less than 100	0-1	Little or no vegetation
Arid	100-400	1-3	Some scrubs, some grassland
Semi-arid	400-600	3-4	Scrubs and bushes, grassland
Sub - humid	600-1200	4-6	Bushes to woodland, grassland
Moist sub-humid	1200-1500	6-9	Forest and woodland
Humid	more than 1500	9-12	Tropical rain forest

The wet period refers to the period during which the rainfall is higher than the evapotranspiration. Evapotranspiration is the sum of the evaporation from the soil surface and the transpiration by the plants.

Desert and Arid Areas: In desert and arid areas irrigation is absolutely essential for crop growth. In these areas the wet period is maximum 3 months. Even when some water is stored in the root zone during the wet period and used later by the plants, it is still not enough to obtain a good harvest. In addition the rainfall is unreliable in desert and arid regions: one year there is a lot of rainfall, the next year there may be only a little. On top of that, the rainfall usually comes in heavy showers resulting in high water losses due to run-off.

Semi Arid Areas: Even in the semi-arid areas irrigation is indispensable for a reliable and good harvest. Some drought resistant crops such as sorghum and millet may give reasonable yields, but also here is a risk of unreliable rainfall and subsequent crop failure.

Sub-humid and Moist Sub-humid Areas: In the sub-humid and moist sub-humid areas, irrigation is required only during the dry season. In principle, on a yearly basis there is a rainfall excess: the yearly rainfall is higher than the yearly evapotranspiration. During the rainy season it may even be too wet for crops like sorghum and millet.

Humid Areas: In the humid areas irrigation is usually not required, except maybe for paddy rice. During more than 9 months per year there is excess rainfall and even in the remaining drier months the plants may use water which has been stored in the root zone during the rainy season.

Major Crop Zones

Crop growth depends not only on rainfall, but also on other climatic factors (most notably sunshine and temperature) and non-climatic factors such as the availability of suitable soils.

During some periods of the year (the cool season) or at high altitudes (i.e. in mountainous areas) crop growth may be prohibited by the low temperature. Other areas may not be suitable for certain crops due to high temperatures.

Following table provides, as an example, data on land use, i.e. the most important crops grown in the 6 major climatic zones of Africa.

Table: Major Climatic Zones and Land use in Africa

Climatic Zone	*Land Use (Major Crops)*
Desert	Hunters and gatherers, nomadic pastoralists, sedentary irrigators around oases, no rainfed agriculture.
Arid	Extensive grazing (nomadic pastoralists), some millet and sorghum under flood irrigation in moist depressions.
Semi-arid	Both nomadic pastoralists and cultivators. Mainly millet and sorghum, also short cycle cowpea, phaseolus beans and groundnuts. No fodder or sown pasture. In cooler parts maize.
Sub-humid	Traditional nomadic pastoralists in dry season and drought years. Crops grown by settlers: millet, sorghum, maize, groundnuts; also cassava, cowpeas, cotton, sweet potatoes, tobacco, rainfed rice, soya bean, mango, cashewnuts. Fodder and sown pasture possible.
Moist sub-humid	Transition zone for agriculture: too wet for seasonal crops, too dry for tree crops. Tropics: maize, cassava; also yams, bananas, pineapple, sugarcane and rice. Winter rainfall areas and East African highlands: wheat and barley.
Humid	Tree crops: oilpalm, rubber, cacao; shifting cultivation based on root crops (yams, cassava, etc.). Also some sorghum, maize, banana, sugarcane, rice. Some tropical hard woods.

Pan Evaporation Method

Evaporation pans provide a measurement of the combined effect of temperature, humidity, windspeed and sunshine on the reference crop evapotranspiration ETo.

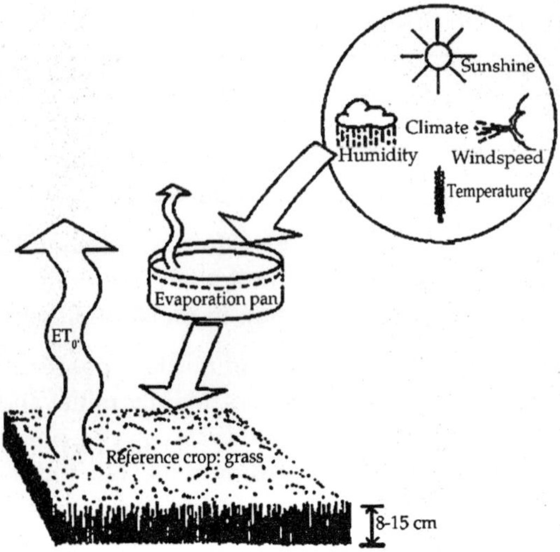

Figure: Pan Evaporation Method

Many different types of evaporation pans are being used. The best known pans are the Class A evaporation pan (circular pan) and the Sunken Colorado pan (square pan).

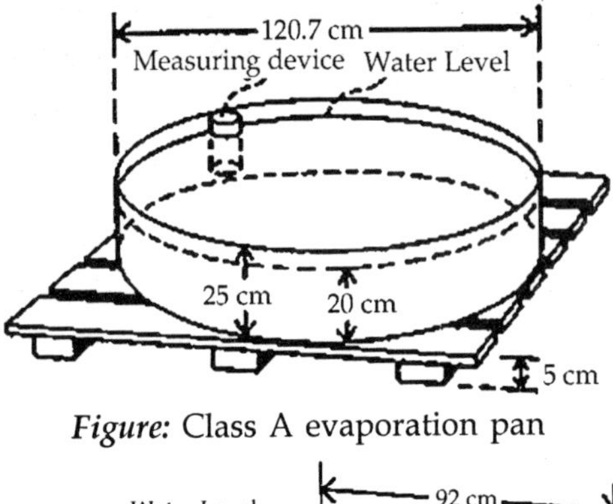

Figure: Class A evaporation pan

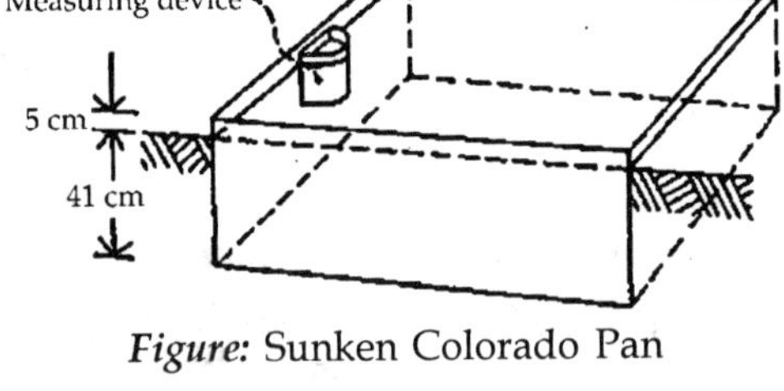

Figure: Sunken Colorado Pan

The principle of the evaporation pan is the following:

- the pan is installed in the field;
- the pan is filled with a known quantity of water (the surface area of the pan is known and the water depth is measured);
- the water is allowed to evaporate during a certain period of time (usually 24 hours). For example, each morning at 7 o'clock a measurement is taken. The rainfall, if any, is measured simultaneously;
- after 24 hours, the remaining quantity of water (i.e. water depth) is measured;
- the amount of evaporation per time unit (the difference between the two measured water depths) is calculated; this is the pan evaporation: E pan (in mm/24 hours);
- the E pan is multiplied by a pan coefficient, K pan, to obtain the ETo.

Formula: ETo = K pan × E pan

with:

ETo: reference crop evapotranspiration

K pan: pan coefficient

E pan: pan evaporation

Figure: Add Water When the Water Depth in the Pan Drops too much.

If the water depth in the pan drops too much (due to lack of rain), water is added and the water depth is measured before and after the water is added. If the water level rises too much (due to rain) water is taken out of the pan and the water depths before and after are measured.

Figure: Take Water Out of the Pan when the Water Depth Rises too Much.

Determination of K Pan

When using the evaporation pan to estimate the ETo, in fact, a comparison is made between the evaporation from the water surface in the pan and the evapotranspiration of the standard grass. Of course the water in the pan and the grass do not react in exactly the same way to the climate. Therefore a special coefficient is used (K pan) to relate one to the other.

The pan coefficient, K pan, depends on:

- the type of pan used;
- the pan environment: if the pan is placed in a fallow or cropped area;
- the climate: the humidity and windspeed.

For the Class A evaporation pan, the K pan varies between 0.35 and 0.85. Average K pan = 0.70.

For the Sunken Colorado pan, the K pan varies between 0.45 and 1.10. Average K pan = 0.80.

The K pan is high if:	The K pan is low if:
the pan is placed in a fallow area	the pan is placed in a cropped area
the humidity is high (i.e. humid)	the humidity is low (i.e. dry)
the windspeed is low	the windspeed is high

Details of the pan coefficient are usually provided by the supplier of the pan.

If the pan factor is not known the average value could be used. These values, however, only refer to the Class A evaporation pan and the Sunken Colorado pan.

Some examples

- Type of pan: Class A evaporation pan

Water depth in pan on day 1 = 150 mm

Water depth in pan on day 2 = 144 mm

Rainfall (during 24 hours) = 0 mm

K pan = 0.75

Formula: ETo = K pan × E pan

Calculation: E pan = 150 - 144 = 6 mm/day

ETo = 0.75 × 6 =4.5 mm/day

* Measuring devices are usually more accurate than the ruler indicated in Figure and thus allow for more accurate readings.

- Type of pan: Sunken Colorado pan

Water depth in pan on day 1 = 411 mm

Water depth in pan on day 2 = 409 mm (after 24 hours)

Rainfall (during 24 hours) = 7 mm

K pan = 0.90

Formula: ETo = K pan × E pan

Calculation: E pan = 411 - 409 + 7 = 9 mm/day

ETo = 0.90 × 9 = 8.1 mm/day

DATA SHEET 1 Determination ETo: Pan Evaporation Method

Blaney-Criddle Method

If no measured data on pan evaporation are available locally, a theoretical method (e.g. the Blaney-Criddle method) to calculate the reference crop evapotranspiration ETo has to be used. There are a large number of theoretical methods to determine the ETo. Many of them have been determined and tested locally. If such local formulae are available they should be used. If such local formulae are not available one of the general theoretical methods has to be used.

The most commonly used theoretical method is the modified Penman method which is described in detail in FAO Irrigation and Drainage Paper 24. This method, however, is rather complicated and beyond the scope of this manual.

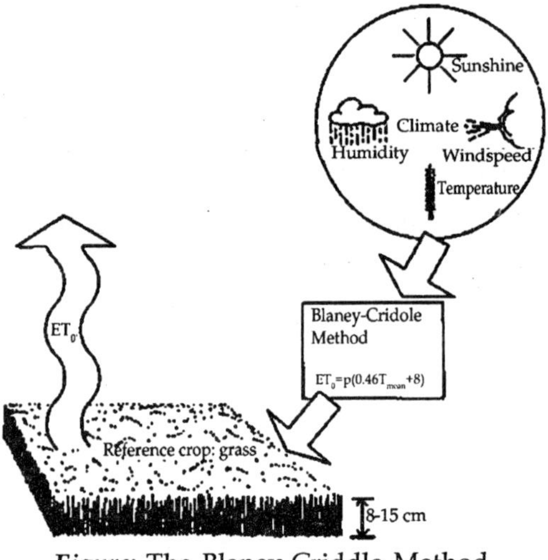

Figure: The Blaney-Criddle Method.

Here only the Blaney-Criddle method is given. The Blaney-Criddle method is simple, using measured data on temperature only. It should be noted, however, that this method is not very accurate; it provides a rough estimate or "order of magnitude" only. Especially under "extreme" climatic conditions the Blaney-Criddle method is inaccurate: in windy, dry, sunny areas, the ETo is underestimated (up to some 60 per cent), while in calm, humid, clouded areas, the ETo is overestimated (up to some 40 per cent).

The Blaney-Criddle formula: ETo = p (0.46 T mean +8)

ETo = Reference crop evapotranspiration (mm/day) as an average for a period of 1 month

T mean = mean daily temperature (°C)

p = mean daily percentage of annual daytime hours

The use of the Blaney-Criddle formula

Step 1: Determination of the mean daily temperature: T mean

The Blaney-Criddle method always refers to mean monthly values, both for the temperature and the ETo. If, for example, it is found that T mean in March is 28°C, it means that during the whole month of March the mean daily temperature is 28°C.

If in a local meteorological station the daily minimum and maximum temperatures are measured, the mean daily temperature is calculated as follows:

$$T_{max} = \frac{\text{sum of all T max values during the month}}{\text{number of days of the month}}$$

$$T_{min} = \frac{\text{sum of all T min values during the month}}{\text{number of days of the month}}$$

$$T_{mean} = \frac{T_{max} + T_{min}}{2}$$

Step 2: Determination of the mean daily percentage of annual daytime hours: p

To determine the value of p. following table is used. To be able to determine the p value it is essential to know the approximate latitude of the area: the number of degrees north or south of the equator.

Suppose the p value for the month March has to be determined for an area with a latitude of 45° South. From following table it can be seen that the p value during March = 0.28.

Table: Mean Daily Percentage (P) of Annual Daytime Hours for Different Latitudes

Latitude	North	Jan	Feb	Mar	Apr	May	June	July	Aug	Sept	Oct	Nov	Dec
	South	July	Aug	Sept	Oct	Nov	Dec	Jan	Feb	Mar	Apr	May	June
60°		.15	.20	.26	.32	.38	.41	.40	.34	.28	.22	.17	.13
55		.17	.21	.26	.32	.36	.39	.38	.33	.28	.23	.18	.16
50		.19	.23	.27	.31	.34	.36	.35	.32	.28	.24	.20	.18
45		.20	.23	.27	.30	.34	.35	.34	.32	.28	.24	.21	.20
40		.22	.24	.27	.30	.32	.34	.33	.31	.28	.25	.22	.21
35		.23	.25	.27	.29	.31	.32	.32	.30	.28	.25	.23	.22
30		.24	.25	.27	.29	.31	.32	.31	.30	.28	.26	.24	.23
25		.24	.26	.27	.29	.30	.31	.31	.29	.28	.26	.25	.24
20		.25	.26	.27	.28	.29	.30	.30	.29	.28	.26	.25	.25
15		.26	.26	.27	.28	.29	.29	.29	.28	.28	.27	.26	.25
10		.26	.27	.27	.28	.28	.29	.29	.28	.28	.27	.26	.26
5		.27	.27	.27	.28	.28	.28	.28	.28	.28	.27	.27	.27
0		.27	.27	.27	.27	.27	.27	.27	.27	.27	.27	.27	.27

Step 3: Calculate ETo, using the formula: ETo = p (0.46 T mean + 8)

For example, when p = 0.29 and T mean = 21.5°C the ETo is calculated as follows:

$$ETo = 0.29\ (0.46 \times 21.5 + 8) = 0.29\ (9.89 + 8) = 0.29 \times 17.89 = 5.2 \text{ mm/day}$$

Calculation Example Blaney-Criddle

It is given that,

Latitude - 35° North

Mean T max. in April = 29.5°C

Mean T min. in April = 19.4°C

We have to determine for the month April the mean ETo in mm/day using the Blaney-Criddle method.

We will get,

Formula: ETo = p (0.46 T mean + 8)

Step 1: determine T mean:

$$T_{mean} = \frac{T_{max} + T_{min}}{2} = \frac{29.5 + 19.4}{2} = 24.5°C$$

Step 2: determine p: Latitude: 35° North

Month: April

From above table: p = 0.29

Step 3: calculate ETo: ETo = 0.29 (0.46 × 24.5 + 8) = 5.6 mm/day

Thus the mean reference crop evapotranspiration ETo = 5.6 mm/day during the whole month of April.

Indicative Values of ETo

If only a rough estimate of the ETo value is required. Following table can be used.

Table: Indicative Values of Eto (MM/Day)

	Mean daily temperature		
Climatic zone	*low (less than 15°C)*	*medium (15-25°C)*	*high (more than 25°C)*
Desert/arid	4-6	7-8	9-10
Semi arid	4-5	6-7	8-9
(Moist) Sub-humid	3-4	5-6	7-8
Humid	1-2	3-4	5-6

Influence of Crop Type on Crop Water Needs (Kc)

Concept and Perception

The influence of the climate is given by the reference crop evapotranspiration ETo; the reference crop used for this purpose is grass.

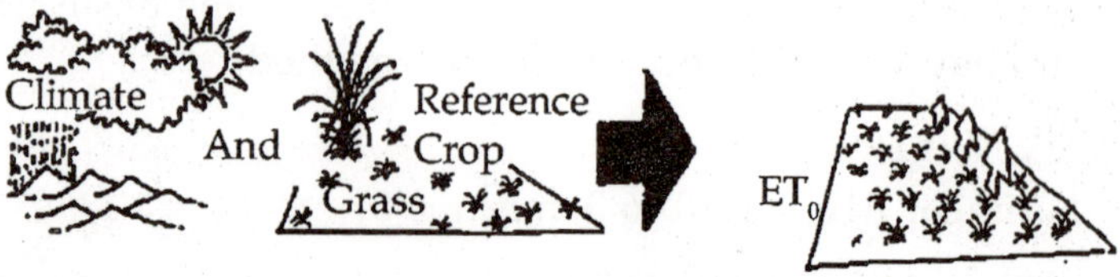

Figure: Reference crop evapotranspiration

This section deals with the influence of the crop type and growth stage on crop water needs. In other words, this section discusses the relationship between the reference grass crop and the crop actually grown in the field.

The relationship between the reference grass crop and the crop actually grown is given by the crop factor, Kc, as shown in the following formula:

$$ETo \times Kc = ET\ crop$$

with ET crop = crop evapotranspiration or crop water need (mm/day)

Kc = crop factor

ETo = reference evapotranspiration (mm/day)

Both ET crop and ETo are expressed in the same unit: usually in mm/day (as an average for a period of one month) or in mm/month.

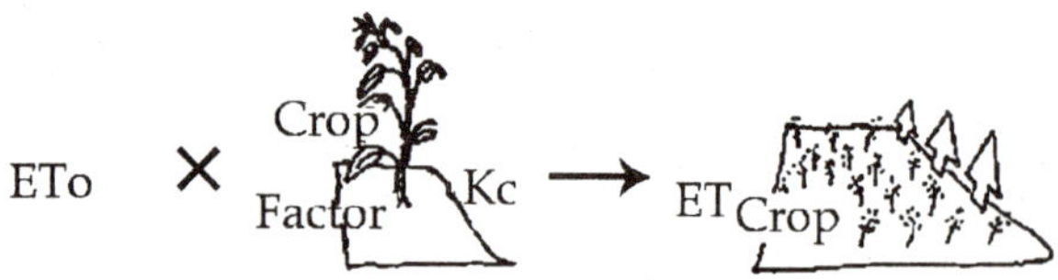

Figure: Crop evapotranspiration or crop water need

The crop factor, Kc, mainly depends on:

- the type of crop;
- the growth stage of the crop;
- the climate.

Kc and the Type of Crop: Fully developed maize, with its large leaf area will be able to transpire, and thus use, more water than the reference grass crop: Kc, maize is higher than 1. Cucumber, also fully developed, will use less water than the reference grass crop: Kc, cucumber is less than 1.

Kc and the Growth Stage of the Crop: A certain crop will use more water once it is fully developed, compared to a crop which has just recently been planted.

Kc and the Climate: The climate influences the duration of the total growing period and the various growth stages. In a cool climate a certain crop will grow slower than in a warm climate.

Thus, to determine the crop factor Kc, it is necessary, for each crop, to know the total length of the growing season and the lengths of the various growth stages.

The determination of the Kc values for the various growth stages of the crops involves several steps:

- *Step 1:* Determination of the total growing period of each crop;
- *Step 2:* Determination of the various growth stages of each crop;
- *Step 3:* Determination of the Kc values for each crop for each of the growth stages.

Determination of the Total Growing Period

The total growing period (in days) is the period from sowing or transplanting to the last day of the harvest. It is mainly dependent on:

- the type of crop and the variety;
- the climate;
- the planting date.

As the growing period heavily depends on local circumstances (e.g. local crop varieties) it is always best to obtain these data locally. Only if no data are available locally should following table be used.

As can be seen from following table there is a large variation of values not only between crops but also within one crop type. In general it can be assumed that the growing period for a certain crop is longer when the climate is cool and shorter when the climate is warm.

Table: Indicative Values of the Total Growing Period

Crop	*Total growing period (days)*	*Crop*	*Total growing period (days)*
Alfalfa	100-365	Millet	105-140
Banana	300-365	Onion green	70-95
Barley/Oats/Wheat	120-150	Onion dry	150-210
Bean green	75-90	Peanut/Groundnut	130-140
Bean dry	95-110	Pea	90-100
Cabbage	120-140	Pepper	120-210
Carrot	100-150	Potato	105-145
Citrus	240-365	Radish	35-45
Cotton	180-195	Rice	90-150
Cucumber	105-130	Sorghum	120-130
Eggplant	130-140	Soya bean	135-150
Flax	150-220	Spinach	60-100
Grain/small	150-165	Squash	95-120
Lentil	150-170	Sugarbeet	160-230
Lettuce	75-140	Sugarcane	270-365
Maize sweet	80-110	Sunflower	125-130
Maize grain	125-180	Tobacco	130-160
Melon	120-160	Tomato	135-180

Determination of the Growth Stages

Once the total growing period is known, the duration (in days) of the various growth stages has to be determined.

The total growing period is divided into 4 growth stages:

1. The initial stage: this is the period from sowing or transplanting until the crop covers about 10 per cent of the ground.
2. The crop development stage: this period starts at the end of the initial stage and lasts until the full ground cover has been reached (ground cover 70-80%); it does not necessarily mean that the crop is at its maximum height.
3. The mid - season stage: this period starts at the end of the crop development stage and lasts until maturity; it includes flowering and grain-setting.
4. The late season stage: this period starts at the end of the mid season stage and lasts until the last day of the harvest; it includes ripening.

Following table shows the duration of the various growth stages for some of the major field crops. For each crop the "minimum" and "maximum" duration of total growing period have been taken and subdivided in the various growth stages.

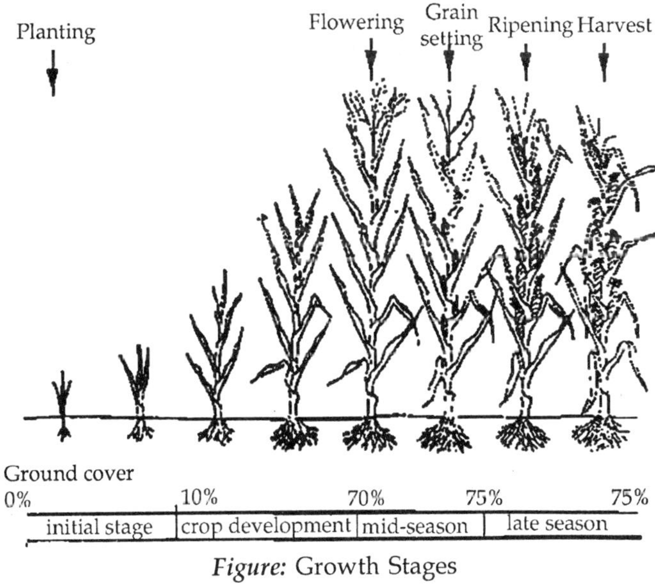

Figure: Growth Stages

Table: Approximate Duration of Growth Stages for Various Field Crops

	Total	Initial stage	Crop Develop-ment stage	Mid season stage	Late season stage
Barley/Oats/Wheat	120	15	25	50	30
	150	15	30	65	40
Bean/green	75	15	25	25	10
	90	20	30	30	10
Bean/dry	95	15	25	35	20
	110	20	30	40	20
Cabbage	120	20	25	60	15
	140	25	30	65	20
Carrot	100	20	30	30	20
	150	25	35	70	20
Cotton/Flax	180	30	50	55	45
	195	30	50	65	50
Cucumber	105	20	30	40	15
	130	25	35	50	20
Eggplant	130	30	40	40	20
	140	30	40	45	25
Grain/small	150	20	30	60	40
	165	25	35	65	40
Lentil	150	20	30	60	40
	170	25	35	70	40
Lettuce	75	20	30	15	10
	140	35	50	45	10
Maize, sweet	80	20	25	25	10
	110	20	30	50	10
Maize, grain	125	20	35	40	30
	180	30	50	60	40
Melon	120	25	35	40	20
	160	30	45	65	20

Contd...

	Total	Initial stage	Crop Development stage	Mid season stage	Late season stage
Millet	105	15	25	40	25
	140	20	30	55	35
Onion/green	70	25	30	10	5
	95	25	40	20	10
Onion/dry	150	15	25	70	40
	210	20	35	110	45
Peanut/Groundnut	130	25	35	45	25
	140	30	40	45	25
Pea	90	15	25	35	15
	100	20	30	35	15
Pepper	120	25	35	40	20
	210	30	40	110	30
Potato	105	25	30	30	20
	145	30	35	50	30
Radish	35	5	10	15	5
	40	10	10	15	5
Sorghum	120	20	30	40	30
	130	20	35	45	30
Soya bean	135	20	30	60	25
	150	20	30	70	30
Spinach	60	20	20	15	5
	100	20	30	40	10
Squash	95	20	30	30	15
	120	25	35	35	25
Sugarbeet	160	25	35	60	40
	230	45	65	80	40
Sunflower	125	20	35	45	25
	130	25	35	45	25
Tomato	135	30	40	40	25
	180	35	45	70	30

Example:

Carrots: the "minimum" growing period is 100 days. This growing period corresponds with the following duration of growth stages:

Initial stage	: 20 days
Crop development stage	: 30 days
Mid-season stage	: 30 days
Late season stage	: 20 days
Total	***100 days***

For the "maximum" growing period of 150 days the following values apply: respectively 25, 35, 70 and 20 days.

Should, under certain local circumstances, the duration of the growing period be 120 days, the duration of the growth stages could be estimated as follows:

Initial stage	: 25 days
Crop development stage	: 35 days
Mid-season stage	: 40 days
Late season stage	: 20 days
Total	***120 days***

1. The table always refers to "sown" crops. When the crop is transplanted, the length of the initial stage should be reduced. For example:

Tomatoes : growing period 180 days from sowing

Direct sowing : initial stage 35 days

Transplanted : (estimated) initial stage 15 days

The growing period from transplant is thus (180 - 20) = 160 days

Direct Sowing:

Initial stage	crop development stage	mid-season stage	late season stage
35 days	45 days	70 days	30 days

Transplanting:

Initial stage	crop development stage	mid-season stage	late season stage
15 days	45 days	70 days	30 days

2. When a crop is harvested "green" or "fresh" the late season stage is short.

   Compare, for example, green beans with dry beans. The duration of the late season stage is 10 and 20 days respectively.

3. If a crop is planted in the winter or is growing in the cool season the total growing period is long. The same is the case with the individual lengths of growing stages. The difference will be most pronounced for the stage during which the temperature is the lowest.

It should be kept in mind that the influence of variations in the total growing period on the crop water need is very important. Less important is the choice of the various lengths of growth stages.

In other words: it is important to obtain (preferably locally) an accurate estimate of the total growing period. The duration of the four growth stages can be estimated.

Note: The sum of the four growth stages should always equal the total growing period.

For examples, estimate the duration of the four growth stages for the following crops: cotton (190 days), lentils (160 days), sweet maize (100 days), potatoes (130 days), tomatoes/transplanted (140 days from transplant), etc. (the figures in brackets refer to the total growing period).

Crop	*Initial stage (days)*	*Crop dev. stage (days)*	*Mid-season stage (days)*	*Late season stage (days)*
Cotton (190 days)	30	50	60	50
Lentils (160 days)	25	30	65	40
Sweet Maize (100 days)	20	30	40	10
Potatoes (130 days)	30	35	40	25

Contd...

Crop	*Initial stage (days)*	*Crop dev. stage (days)*	*Mid-season stage (days)*	*Late season stage (days)*
Tomatoes/ Transplanted (140 days from transplant)	15	40	60	25

Surface irrigation of a large field

Determination of Crop Factors

Per crop, four crop factors have to be determined: one crop factor for each of the four growth stages. Following table indicates per crop the Kc values for each of the four growth stages.

Table: Values of the Crop Factor (Kc) for Various Crops and Growth Stages

Crop	*Initial stage*	*Crop dev. stage*	*Mid-season stage*	*Late season stage*
Barley/Oats/Wheat	0.35	0.75	1.15	0.45
Bean, green	0.35	0.70	1.10	0.90
Bean, dry	0.35	0.70	1.10	0.30
Cabbage/Carrot	0.45	0.75	1.05	0.90
Cotton/Flax	0.45	0.75	1.15	0.75
Cucumber/Squash	0.45	0.70	0.90	0.75
Eggplant/Tomato	0.45	0.75	1.15	0.80
Grain/small	0.35	0.75	1.10	0.65
Lentil/Pulses	0.45	0.75	1.10	0.50
Lettuce/Spinach	0.45	0.60	1.00	0.90
Maize, sweet	0.40	0.80	1.15	1.00
Maize, grain	0.40	0.80	1.15	0.70
Melon	0.45	0.75	1.00	0.75
Millet	0.35	0.70	1.10	0.65
Onion, green	0.50	0.70	1.00	1.00
Onion, dry	0.50	0.75	1.05	0.85
Peanut/Groundnut	0.45	0.75	1.05	0.70

Contd...

Crop	*Initial stage*	*Crop dev. stage*	*Mid-season stage*	*Late season stage*
Pea, fresh	0.45	0.80	1.15	1.05
Pepper, fresh	0.35	0.70	1.05	0.90
Potato	0.45	0.75	1.15	0.85
Radish	0.45	0.60	0.90	0.90
Sorghum	0.35	0.75	1.10	0.65
Soya bean	0.35	0.75	1.10	0.60
Sugarbeet	0.45	0.80	1.15	0.80
Sunflower	0.35	0.75	1.15	0.55
Tobacco	0.35	0.75	1.10	0.90

The table above shows average Kc values for the various crops and growth stages. In fact, the Kc is also dependent on the climate and, in particular, on the relative humidity and the windspeed. The values indicated above should be reduced by 0.05 if the relative humidity is high (RH > 80%) and the windspeed is low (u < 2 m/sec), e.g. Kc = 1.15 becomes Kc = 1.10. The values should be increased by 0.05 if the relative humidity is low (RH < 50%) and the windspeed is high (u > 5 m/sec), e.g. Kc = 1.05 becomes Kc = 1.10.

Irrigation Process

Irrigation is an artificial application of water to the soil. It is usually used to assist in growing crops in dry areas and during periods of inadequate rainfall. Additionally, irrigation also has a few other uses in crop production, which include protecting plants against frost, suppressing weed growing in rice fields and helping in preventing soil consolidation. In contrast, agriculture that relies only on direct rainfall is referred to as rain-fed farming. Irrigation is often studied together with drainage, which is the natural or artificial removal of surface and subsurface water from a given area.

Irrigation is also a term used in medical/dental fields to refer to flushing and washing out anything with water or another liquid.

Historical Backdrop

Archaeological investigation has identified evidence of irrigation in Mesopotamia and Egypt as far back as the 6th millennium BCE, where barley was grown in areas where the natural rainfall was insufficient to support such a crop.

In the Zana Valley of the Andes Mountains in Peru, archaeologists found remains of three irrigation canals radiocarbon dated from the 4th millennium BCE, the 3rd millennium BCE and the 9th century CE. These canals are the earliest record of irrigation in the New World. Traces of a canal possibly dating from the 5th millennium BCE were found under the 4th millennium canal. Sophisticated irrigation and storage systems were developed by the Indus Valley Civilization in Pakistan and North India, including the reservoirs at Girnar in 3000 BCE and an early canal irrigation system from circa 2600 BCE. Large scale agriculture was practiced and an extensive network of canals was used for the purpose of irrigation.

There is evidence of the ancient Egyptian pharaoh Amenemhet III in the twelfth dynasty (about 1800 BCE) using the natural lake of the Faiyum Oasis as a reservoir to store surpluses of water for use during the dry seasons, as the lake swelled annually as caused by the annual flooding of the Nile.

The Qanats, developed in ancient Persia in about 800 BCE, are among the oldest known irrigation methods still in use today. They are now found in Asia, the middle east and north Africa. The system comprises a network of vertical wells and gently sloping tunnels driven into the sides of cliffs and steep hills to tap groundwater. The noria, a water wheel with clay pots around the rim powered by the flow of the stream (or by animals where the water source was still), was first brought into use at about this time, by Roman settlers in North Africa. By 150 BCE the pots were fitted with valves to allow smoother filling as they were forced into the water.

The irrigation works of ancient Sri Lanka, the earliest dating from about 300 BCE, in the reign of King Pandukabhaya and under continuous development for the next thousand years, were one of the most complex irrigation systems of the ancient world. In addition to underground canals, the Sinhalese were the first to build completely artificial reservoirs to store water. The system was extensively restored and further extended during the reign of King Parakrama Bahu (1153-1186 CE).

The oldest known hydraulic engineers of China were Sunshu Ao (6th century BCE) of the Spring and Autumn Period and Ximen Bao (5th century BCE) of the Warring States period, both of whom worked on large irrigation projects. In the Szechwan region belonging to the State of Qin of ancient China, the Dujiangyan Irrigation System was built in 256 BCE to irrigate an enormous area of farmland that today still supplies water. By the 2nd century AD, during the Han Dynasty, the Chinese also used chain pumps that lifted water from lower elevation to higher elevation. These were powered by manual foot pedal, hydraulic waterwheels, or rotating mechanical wheels pulled by oxen. The water was used for public works of providing water for urban residential quarters and palace gardens, but mostly for irrigation of farmland canals and channels in the fields.

In fifteenth century Korea the world's first water gauge, *uryanggye*, was discovered in 1441 CE. The inventor was Jang Yeong-sil, a Korean engineer of the Joseon Dynasty, under the active direction of the King, Sejong the Great. It was installed in irrigation tanks as part of a nationwide system to measure and collect rainfall for agricultural applications. With this instrument, planners and farmers could make better use of the information gathered in the survey.

Present Extent

In the middle of the 20th century, the advent of diesel and electric motors led for the first time to systems that could pump groundwater out of major aquifers faster than it was recharged. This can lead to permanent loss of aquifer capacity, decreased water quality, ground subsidence, and other problems. The future of food production in such areas as the North China Plain, the Punjab, and the Great Plains of the US is threatened.

At the global scale 2,788,000 sq km (689 million acres) of agricultural land was equipped with irrigation infrastructure around the year 2000. About 68% of the area equipped for irrigation is located in Asia, 17% in America, 9% in Europe, 5% in Africa and 1% in Oceania. The largest contiguous areas of high irrigation density are found in North India and Pakistan along the rivers Ganges and Indus, in the Hai He, Huang He and Yangtze basins in China, along the Nile river in Egypt and Sudan, in the Mississippi-Missouri river basin and in parts of California. Smaller irrigation areas are spread across almost all populated parts of the world.

Types of Irrigation

Various types of irrigation techniques differ in how the water obtained from the source is distributed within the field. In general, the goal is to supply the entire field uniformly with water, so that each plant has the amount of water it needs, neither too much nor too little.

Surface Irrigation: In surface irrigation systems water moves over and across the land by simple gravity flow in order to wet it and to infiltrate into the soil. Surface irrigation can be subdivided into furrow, borderstrip or basin irrigation. It is often called flood irrigation when the irrigation results in flooding or near flooding of the cultivated land. Historically, this has been the most common method of irrigating agricultural land.

Where water levels from the irrigation source permit, the levels are controlled by dikes, usually plugged by soil. This is often seen in terraced rice fields (rice paddies), where the method is used to flood or control the level of water in each distinct field. In some cases, the water is pumped, or lifted by human or animal power to the level of the land.

Localised Irrigation: Localised irrigation is a system where water is distributed under low pressure through a piped network, in a pre-determined pattern, and applied as a small discharge to each plant or adjacent to it. Drip irrigation, spray or micro-sprinkler irrigation and bubbler irrigation belong to this category of irrigation methods.

Drip Irrigation: Drip irrigation, also known as trickle irrigation, functions as its name suggests. Water is delivered at or near the root zone of plants, drop by drop. This method can be the most water-efficient method of irrigation, if managed properly, since evaporation and run-off are minimised. In modern agriculture, drip irrigation is often combined with plastic mulch, further reducing evaporation, and is also the means of delivery of fertilizer. The process is known as *fertigation.*

Deep percolation, where water moves below the root zone, can occur if a drip system is operated for too long of a duration or if the delivery rate is too high. Drip irrigation methods range from very high-tech and computerised to low-tech and labour-intensive. Lower water pressures are usually needed than for most other types of systems, with the exception of low energy centre pivot systems and surface irrigation systems, and the system can be designed for uniformity throughout a field or for precise water delivery to individual plants in a landscape containing a mix of plant species. Although it is difficult to regulate pressure on steep slopes, pressure compensating emitters are available, so the field does not have to be level.

High-tech solutions involve precisely calibrated emitters located along lines of tubing that extend from a computerised set of valves. Both pressure regulation and filtration to remove particles are important. The tubes are usually black (or buried under soil or mulch) to prevent the growth of algae and to protect the polyethylene from degradation due to ultraviolet light. But drip irrigation can also be as low-tech as a porous clay vessel sunk into the soil and occasionally filled from a hose or bucket. Subsurface drip irrigation has been used successfully on lawns, but it is more expensive than a more traditional sprinkler system. Surface drip systems are not cost-effective (or aesthetically pleasing) for lawns and golf courses. In the past one of the main disadvantages of the subsurface drip irrigation (SDI) systems, when used for turf, was the fact of having to install the plastic lines very close to each other in the ground, therefore disrupting the turfgrass area. Recent technology developments on drip installers like the drip installer at New Mexico State University Arrow Head Centre, places the line underground and covers the slit leaving no soil exposed.

Sprinkler Irrigation: In sprinkler or overhead irrigation, water is piped to one or more central locations within the field and distributed by overhead high-pressure sprinklers or guns. A system utilising sprinklers, sprays, or guns mounted overhead on permanently installed risers is often referred to as a *solid-set* irrigation system. Higher pressure sprinklers that rotate are called *rotors* and are driven by a ball drive, gear drive, or impact mechanism. Rotors can be designed to rotate in a full or partial circle. Guns are similar to rotors, except that they generally operate at very high pressures of 40 to 130 lbf/sq in (275 to 900 kPa) and flows of 50 to 1200 US gal/min (3 to 76 L/s), usually with nozzle diameters in the range of 0.5 to 1.9 inches (10 to 50 mm). Guns are used not only for irrigation, but also for industrial applications such as dust suppression and logging.

Sprinklers may also be mounted on moving platforms connected to the water source by a hose. Automatically moving wheeled systems known as *travelling sprinklers* may irrigate areas such as small farms, sports fields, parks, pastures, and cemeteries unattended. Most of these utilise a length of polyethylene tubing wound on a steel drum. As the tubing is wound on the drum powered by the irrigation water or a small gas engine, the sprinkler is pulled across the field.

When the sprinkler arrives back at the reel the system shuts off. This type of system is known to most people as a "water-reel" travelling irrigation sprinkler and they are used extensively for dust suppression, irrigation, and land application of waste water. Other travellers use a flat rubber hose that is dragged along behind while the sprinkler platform is pulled by a cable. These cable-type travellers are definitely old technology and their use is limited in today's modern irrigation projects.

Centre Pivot Irrigation: Centre pivot irrigation is a form of sprinkler irrigation consisting of several segments of pipe (usually galvanised steel or aluminium) joined together and supported by trusses, mounted on wheeled towers with sprinklers positioned along its length. The system moves in a circular pattern and is fed with water from the pivot point at the centre of the arc. These systems are common in parts of the United States where terrain is flat.

Most centre pivot systems now have drops hanging from a u-shaped pipe called a *gooseneck* attached at the top of the pipe with sprinkler heads that are positioned a few feet (at most) above the crop, thus limiting evaporative losses. Drops can also be used with drag hoses or bubblers that deposit the water directly on the ground between crops. The crops are planted in a circle to conform to the centre pivot. This type of system is known as LEPA (Low Energy Precision Application). Originally, most centre pivots were water powered. These were replaced by hydraulic systems (*T-L Irrigation*) and electric motor driven systems (*Lindsay, Reinke, Valley, Zimmatic, Pierce, Grupo Chamartin*. Most systems today are driven by an electric motor mounted low on each span. This drives a reduction gearbox and transverse driveshafts transmit power to another reduction gearbox mounted behind each wheel. Precision controls, some with GPS location and remote computer monitoring, are now available.

Lateral Move (Side Roll, Wheel Line) Irrigation: A series of pipes, each with a wheel of about 1.5 m diameter permanently affixed to its midpoint and sprinklers along its length, are coupled together at one edge of a field. Water is supplied at one end using a large hose. After sufficient water has been applied, the hose is removed and the remaining assembly rotated either by hand or with a purpose-built mechanism, so that the sprinklers move 10 m across the field. The hose is reconnected. The process is repeated until the opposite edge of the field is reached. This system is less expensive to install than a centre pivot, but much more labour intensive to operate, and it is limited in the amount of water it can carry. Most systems utilise 4 or 5-inch (130 mm) diameter aluminium pipe. One feature of a lateral move system is that it consists of sections that can be easily disconnected. They are most often used for small or oddly-shaped fields, such as those found in hilly or mountainous regions, or in regions where labour is inexpensive.

Subirrigation: Subirrigation also sometimes called *seepage irrigation* has been used for many years in field crops in areas with high water tables. It is a method of artificially raising the water table to allow the soil to be moistened from below the plants' root zone. Often those systems are located on permanent grasslands in lowlands or river valleys and combined with drainage infrastructure. A system of pumping stations, canals, weirs and gates allows it to increase or decrease the water level in a network of ditches and thereby control the water table.

Sub-irrigation is also used in commercial greenhouse production, usually for potted plants. Water is delivered from below, absorbed upwards, and the excess collected for recycling. Typically, a solution of water and nutrients floods a container or flows through a trough for a short period of time, 10-20 minutes, and is then pumped back into a holding tank for reuse. Sub-irrigation in greenhouses requires fairly sophisticated, expensive equipment and management. Advantages are water and nutrient conservation, and labour-saving through lowered system maintenance and automation. It is similar in principle and action to subsurface drip irrigation.

Manual Irrigation using Buckets or Watering Cans

These systems have low requirements for infrastructure and technical equipment but need high labour inputs. Irrigation using watering cans is to be found for example in peri-urban agriculture around large cities in some African countries.

Automatic, Non-electric Irrigation using Buckets and Ropes

Besides the common manual watering by bucket, an automated, natural version of this also exist. Using plain polyester ropes combined with a prepared ground mixture can be used to water plants from a vessel filled with water. The ground mixture would need to be made depending on the plant itself, yet would mostly consist of black potting soil, vermiculite and perlite. This system would (with certain crops) allow you to save expenses as it does not consume any electricity and only little water (unlike sprinklers, water timers, ...). However, it may only be used with certain crops (probably mostly larger crops that do not need a humid environment; perhaps, e.g. paprika's).

Irrigation using Stones to Catch Water from Humid Air

In countries where at night, humid air sweeps the countryside, stones are used to catch water from the humid air by condensation. This is for example practiced in the vineyards at Lanzarote.

Dry Terraces for Irrigation and Water Distribution

In subtropical countries as Mali and Senegal, a special type of terracing (without flood irrigation or intent to flatten farming ground) is used. Here, a 'stairs' is made through the use of ground level differences which helps to decrease water evaporation and also distributes the water to all patches (sort of irrigation).

Sources of Irrigation Water

Sources of irrigation water can be groundwater extracted from springs or by using wells, surface water withdrawn from rivers, lakes or reservoirs or non-conventional sources like treated wastewater, desalinated water or drainage water. A special form of irrigation using surface water is spate irrigation, also called floodwater harvesting. In case of a flood (spate) water is diverted to normally dry river beds (wadi's) using a network of dams, gates and channels and spread over large areas. The moisture stored in the soil will be used thereafter to grow crops. Spate irrigation areas are in particular located in semi-arid or arid, mountainous regions.

While floodwater harvesting belongs to the accepted irrigation methods, rainwater harvesting is usually not considered as a form of irrigation. Rainwater harvesting is the collection of run-off water from roofs or unused land and the concentration of this water on cultivated land. Therefore, this method is considered as a water concentration method.

How an In-ground Irrigation System Works ?

Most commercial and residential irrigation systems are "in ground" systems, which means that everything is buried in the ground. With the pipes, sprinklers, and irrigation valves being hidden, it makes for a cleaner, more presentable landscape without garden hoses or other items having to be moved around manually.

Water Source and Piping

The beginning of a sprinkler system is the water source. This is usually a tap into an existing (city) water line or a pump that pulls water out of a well or a pond. The water travels through pipes from the water source through the valves to the sprinklers. The pipes from the water source up to the irrigation valves are called "mainlines", and the lines from the valves to the sprinklers are called "lateral lines". Most piping used in irrigation systems today are HDPE and MDPE or PVC or PEX plastic pressure pipes due to their ease of installation and resistance to corrosion. After the water source, the water usually travels through a check valve. This prevents water in the irrigation lines from being pulled back into and contaminating the clean water supply.

Controllers, Zones and Valves

Most Irrigation systems are divided into zones. A zone is a single Irrigation Valve and one or a group of sprinklers that are connected by pipes. Irrigation Systems are divided into zones because there is usually not enough pressure and available flow to run sprinklers for an entire yard or sports field at once. Each zone has a solenoid valve on it that is controlled via wire by an Irrigation Controller.

The Irrigation Controller is either a mechanical or electrical device that signals a zone to turn on at a specific time and keeps it on for a specified amount of time. "Smart Controller" is a recent term used to describe a controller that is capable of adjusting the watering time by itself in response to current environmental conditions. The smart controller determines current conditions by means of historic weather data for the local area, a moisture sensor (water potential or water content), weather station, or a combination of these.

Sprinklers

When a zone comes on, the water flows through the lateral lines and ultimately ends up at the irrigation Sprinkler heads. Most sprinklers have pipe thread inlets on the bottom of them which allows a fitting and the pipe to be attached to them. The sprinklers are usually installed with the top of the head flush with the ground surface. When the water is pressurised, the head will pop up out of the ground and water the desired area until the valve closes and shuts off that zone. Once there is no more water pressure in the lateral line, the sprinkler head will retract back into the ground.

Problems in Irrigation

- Competition for surface water rights.
- Depletion of underground aquifers.
- Ground subsidence (e.g. New Orleans, Louisiana)
- Underirrigation gives poor soil salinity control which leads to increased soil salinity with consequent build up of toxic salts on soil surface in areas with high evaporation. This requires either leaching to remove these salts and a method of drainage to carry the salts away.

- Overirrigation because of poor distribution uniformity or management wastes water, chemicals, and may lead to water pollution.
- Deep drainage (from over-irrigation) may result in rising water tables which in some instances will lead to problems of irrigation salinity.
- Irrigation with saline or high-sodium water may damage soil structure.

Surface Irrigation

Surface irrigation is defined as the group of application techniques where water is applied and distributed over the soil surface by gravity. It is by far the most common form of irrigation throughout the world and has been practiced in many areas virtually unchanged for thousands of years.

Surface irrigation is often referred to as flood irrigation, implying that the water distribution is uncontrolled and therefore, inherently inefficient. In reality, some of the irrigation practices grouped under this name involve a significant degree of management (for example surge irrigation). Surface irrigation comes in three major types; level basin, furrow and border strip.

Basin Irrigation

Level basin irrigation has historically been used in small areas having level surfaces that are surrounded by earth banks. The water is applied rapidly to the entire basin and is allowed to infiltrate. Basins may be linked sequentially so that drainage from one basin is diverted into the next once the desired soil water deficit is satisfied. A "closed" type basin is one where no water is drained from the basin. Basin irrigation is favoured in soils with relatively low infiltration rates (Walker and Skogerboe 1987). Fields are typically set up to follow the natural contours of the land but the introduction of laser levelling and land grading has permitted the construction of large rectangular basins that are more appropriate for mechanised broadacre cropping.

Furrow Irrigation

Furrow irrigation is conducted by creating small parallel channels along the field length in the direction of predominant slope. Water is applied to the top end of each furrow and flows down the field under the influence of gravity. Water may be supplied using gated pipe, siphon and head ditch or bankless systems. The speed of water movement is determined by many factors such as slope, surface roughness and furrow shape but most importantly by the inflow rate and soil infiltration rate. The spacing between adjacent furrows is governed by the crop species, common spacings typically range from 0.75 to 2 metres.

The crop is planted on the ridge between furrows which may contain a single row of plants or several rows in the case of a bed type system. Furrows may range anywhere from less than 100 m to 2000 m long depending on the soil type, location and crop type. Shorter furrows are commonly associated with higher uniformity of application but result in increasing potential for run-off losses. Furrow irrigation is particularly suited to broad-acre row crops such as cotton, maize and sugarcane. It is also practiced in various horticultural industries such as citrus, stone-fruit and tomatoes.

The water can take a considerable period of time to reach the other end, meaning water has been infiltrating for a longer period of time at the top end of the field. This results in poor uniformity with high application at the top end with lower application at the bottom end. In most cases the performance of furrow irrigation can be improved through increasing the speed at which water moves along the field (the advance rate). This can be achieved through increasing flow rates or through the practice of surge irrigation. Increasing the advance rate not only improves the uniformity but also reduces the total volume of water required to complete the irrigation.

Surge Irrigation

Surge Irrigation is a variant of furrow irrigation where the water supply is pulsed on and off in planned time periods (e.g. on for ½ hour off for ½ hour). The wetting and drying cycles reduce infiltration rates resulting in faster advance rates and higher uniformities than continuous flow.

The reduction in infiltration is a result of surface consolidation, filling of cracks and micro pores and the disintegration of soil particles during rapid wetting and consequent surface sealing during each drying phase. The effectiveness of surge irrigation is soil type dependent, for example many clay soils experience a rapid sealing behaviour under continuous flow therefore surge offers little benefit.

Bay or Border Strip Irrigation

Border strip or bay irrigation could be considered as a hybrid of level basin and furrow irrigation. The borders of the irrigated strip are longer and the strips are narrower than for basin irrigation and are orientated to align lengthwise with the slope of the field. The water is applied to the top end of the bay, which is usually constructed to facilitate free-flowing conditions at the downstream end. One common use of this technique includes the irrigation of pasture for dairy production.

Drainage after Harvest or in Rainy Season

Drainage of flooded banks or drainage of extremely wet soil during the rainy season may be done by ditches. Drainage by ditches may be done with crops that require the soil to be wet but not completely saturated (and sometimes, especially not at certain times of year). An example is blueberries. In the rainy season/winter, they require drier soil.

Drip Irrigation

Drip irrigation, also known as trickle irrigation or microirrigation is an irrigation method which minimises the use of water and fertilizer by allowing water to drip slowly to the roots of plants, either onto the soil surface or directly onto the root zone, through a network of valves, pipes, tubing, and emitters.

Modern drip irrigation has arguably become the world's most valued innovation in agriculture since the invention of the impact sprinkler in the 1930s, which replaced flood irrigation. Drip irrigation may also use devices called micro-spray heads, which spray water in a small area, instead of dripping emitters. These are generally used on tree and vine crops with wider root zones. Subsurface drip irrigation (SDI) uses permanently or temporarily buried dripperline or drip tape located at or below the plant roots. It is becoming popular for row crop irrigation, especially in areas where water supplies are limited or recycled water is used for irrigation. Careful study of all the relevant factors like land topography, soil, water, crop and agro-climatic conditions are needed to determine the most suitable drip irrigation system and components to be used in a specific installation.

Historical Backdrop

Drip irrigation has been used since ancient times when buried clay pots were filled with water, which would gradually seep into the grass. Modern drip irrigation began its development in Afghanistan in 1866 when researchers began experimenting with irrigation using clay pipe to create combination irrigation and drainage systems.

In 1913, E.B. House at Colorado State University succeeded in applying water to the root zone of plants without raising the water table. Perforated pipe was introduced in Germany in the 1920s and in 1934, O.E. Nobey experimented with irrigating through porous canvas hose at Michigan State University.

With the advent of modern plastics during and after World War II, major improvements in drip irrigation became possible. Plastic microtubing and various types of emitters began to be used in the greenhouses of Europe and the United States.

The modern technology of drip irrigation was invented in Israel by Simcha Blass and his son Yeshayahu. Instead of releasing water through tiny holes, blocked easily by tiny particles, water was released through larger and longer passageways by using velocity to slow water inside a plastic emitter. The first experimental system of this type was established in 1959 when Blass partnered with Kibbutz Hatzerim to create an irrigation company called Netafim.

Together they developed and patented the first practical surface drip irrigation emitter. This method was very successful and subsequently spread to Australia, North America, and South America by the late-1960s.

In the United States, in the early-1960s, the first drip tape, called *Dew Hose*, was developed by Richard Chapin of Chapin Watermatics (first system established during 1964). Beginning in 1989, Jain irrigation helped pioneer effective water-management through drip irrigation in India. Jain irrigation also introduced some drip irrigation marketing approaches to Indian agriculture such as 'Integrated System Approach', One-Stop-Shop for Farmers, 'Infrastructure Status to Drip Irrigation & Farm as Industry.' The latest developments in the field involve even further reduction in drip rates being delivered and less tendency to clog.

Components and Operation

Components (listed in order from water source):

- Pump or pressurised water source;
- Water Filter(s) - Filtration Systems: Sand Separator like Hydro-Cyclone, Screen Filter, Media Filters;
- Fertigation Systems (Venturi injector) and Chemigation Equipment (optional);
- Backwash Controller;
- Main Line (larger diameter Pipe and Pipe Fittings);
- Hand-operated, electronic, or hydraulic Control Valves and Safety Valves;
- Smaller diameter polytube (often referred to as "laterals");
- Poly fittings and Accessories (to make connections);

- Emitting Devices at plants (ex. Emitter or Drippers, micro spray heads, inline drippers, trickle rings);
- Note that in Drip irrigation systems Pump and valves may be manually or automatically operated by a controller.

Most large drip irrigation systems employ some type of filter to prevent clogging of the small emitter flow path by small waterborne particles. New technologies are now being offered that minimise clogging. Some residential systems are installed without additional filters since potable water is already filtered at the water treatment plant.

Virtually all drip irrigation equipment manufacturers recommend that filters be employed and generally will not honour warranties unless this is done. Last line filters just before the final delivery pipe are strongly recommended in addition to any other filtration system due to fine particle settlement and accidental insertion of particles in the intermediate lines.

Drip and subsurface drip irrigation is used almost exclusively when using recycled municipal waste water. Regulations typically do not permit spraying water through the air that has not been fully treated to potable water standards.

Because of the way the water is applied in a drip system, traditional surface applications of timed-release fertilizer are sometimes ineffective, so drip systems often mix liquid fertilizer with the irrigation water. This is called fertigation; fertigation and chemigation (application of pesticides and other chemicals to periodically clean out the system, such as chlorine or sulphuric acid) use chemical injector such as diaphragm pumps, piston pumps, or venturi pumps. The chemicals may be added constantly whenever the system is irrigating or at intervals. Fertilizer savings of up to 95% are being reported from recent university field tests using drip fertigation and slow water delivery as compared to timed-release and irrigation by micro spray heads.

If properly designed, installed, and managed, drip irrigation may help achieve water conservation by reducing evaporation and deep drainage when compared to other types of irrigation such as flood or overhead sprinklers since water can be more precisely applied to the plant roots. In addition, drip can eliminate many diseases that are spread through water contact with the foliage. Finally, in regions where water supplies are severely limited, there may be no actual water savings, but rather simply an increase in production while using the same amount of water as before. In very arid regions or on sandy soils, the preferred method is to apply the irrigation water as slowly as possible.

Pulsed irrigation is sometimes used to decrease the amount of water delivered to the plant at any one time, thus reducing run-off or deep percolation. Pulsed systems are typically expensive and require extensive maintenance. Therefore, the latest efforts by emitter manufacturers are focused towards developing new technologies that deliver irrigation water at ultra-low flow rates, i.e. less than 1.0 litre per hour. Slow and even delivery further improves water use efficiency without incurring the expense and complexity of pulsed delivery equipment.

Drip irrigation is used by farms, commercial greenhouses, and residential gardeners.

Drip irrigation is adopted extensively in areas of acute water scarcity and especially for crops such as coconuts, containerised landscape trees, grapes, bananas, ber, brinjal, citrus, strawberries, sugarcane, cotton, maize, and tomatoes.

Garden drip irrigation kits are increasingly popular for the homeowner and consist of a timer, hose and emitter.

Advantage and Disadvantages of Drip Irrigation

The advantages of drip irrigation are:

- Minimised fertilizer/nutrient loss due to localised application and reduced leaching.

- High water application efficiency.
- Levelling of the field not necessary.
- Ability to irrigate irregular shaped fields.
- Allows safe use of recycled water.
- Moisture within the root zone can be maintained at field capacity.
- Soil type plays less important role in frequency of irrigation.
- Minimised soil erosion.
- Highly uniform distribution of water, i.e., controlled by output of each nozzle.
- Lower labour cost.
- Variation in supply can be regulated by regulating the valves and drippers.
- Fertigation can easily be included with minimal waste of fertilizers.
- Foliage remains dry thus reducing the risk of disease.
- Usually operated at lower pressure than other types of pressurised irrigation, reducing energy costs.

The disadvantages of drip irrigation are:

- *Expense:* Initial cost can be more than overhead systems.
- *Waste:* The sun can affect the tubes used for drip irrigation, shortening their usable life. Longevity is variable.
- *Clogging:* If the water is not properly filtered and the equipment not properly maintained, it can result in clogging.
- Drip irrigation might be unsatisfactory if herbicides or top dressed fertilizers need sprinkler irrigation for activation.
- Drip tape causes extra cleanup costs after harvest. You'll need to plan for drip tape winding, disposal, recycling or reuse.

- Waste of water, time & harvest, if not installed properly. These systems requires careful study of all the relevant factors like land topography, soil, water, crop and agro-climatic conditions, and suitability of drip irrigation system and its components.
- *Germination Problems:* In lighter soils subsurface drip may be unable to wet the soil surface for germination. Requires careful consideration of the installation depth.

Dripperline

A dripperline is a type of drip irrigation tubing with emitters pre-installed at the factory.

Emitter

An emitter is also called a dripper and is used to transfer water from a pipe or tube to the area that is to be irrigated. Typical emitter flow rates are from 0.16 to 4.0 US gallons per hour (0.6 to 16 L/h). In many emitters, flow will vary with pressure, while some emitters are *pressure compensating*. These emitters employ silicone diaphragms or other means to allow them to maintain a near-constant flow over a range of pressures, for example from 10 to 50 psi (70 to 350 kPa).

Some of the more notable manufacturers of emitter are:

- Chapin Watermatics (Water Town, NY, USA);
- DRTS (USA, Italy, Turkey);
- Eurodrip;
- Jain irrigation Systems (Jalgaon, India);
- Netafim Irrigation;
- Toro Micro-Irrigation;
- Premier Irrigation Adritec Ltd., Kolkata, (Bhanu).

Subirrigation

In agriculture, subirrigation, also known as *seepage irrigation,* is a method of irrigation where water is delivered to the plant root zone from below the soil surface and absorbed upwards. The excess may be collected for reuse.

Subirrigation is used in growing field crops such as tomatoes, peppers, and sugarcane in areas with high water tables such as Florida and in commercial greenhouse operations.

Three basic types of subirrigation system are in general use for potted plants in greenhouses: ebb-and-flow (bench-mounted enclosures holding pots are filled and then drained); trough (water is flowed through bench-mounted, slightly sloping enclosures containing pots); and flooded floor (special sloped concrete flooring is flooded and drained).

Greenhouse subirrigation has been growing in popularity since the 1990s. Advantages are water and nutrient conservation, and labour-saving. The outfitting cost is relatively high. Potential problems, such as the possibility of increased presence of disease in recycle water, have only begun to be investigated.

Controller Irrigation

An irrigation controller is a device to operate automatic irrigation systems such as lawn sprinkler systems and drip irrigation systems. Most controllers have a means of setting the frequency of irrigation, the start time, and the duration of watering. Some controllers have additional features such as multiple programmes to allow different watering frequencies for different types of plants, rain delay settings, input terminals for sensors such as rain and freeze sensors, soil moisture sensors, weather data, remote operation, etc.

There are two basic types of controllers, electric and hydraulic. Most automatic irrigation valves are diaphragm valves, in which the water above the diaphragm must be discharged for the valve to open. In a hydraulic system, the controller and valves are connected via small plastic tubes approximately 4 mm (¼ in) in diameter. The controller opens the tube connected to the valve, allowing that valve to open.

Most newer systems employ electric controllers. In this scenario, the controller is connected to an electrical circuit that operates a solenoid attached to the valves. When the solenoid is actuated, the water above the diaphragm is relieved and the valve opens.

Although sophisticated controllers that allow irrigation schedules to be automatically adjusted according to the weather have been available for many years, until recently these controllers were out of reach of the average consumer. One type is evapotranspiration controllers or "ET controllers". Several manufacturers are now producing controllers that can be automatically updated by either a simple weather sensor, via a pager that receives a daily update from a network of local weather stations, or through soil moisture sensors. One company has also introduced a product that gathers information from the internet to update the watering schedule. Since approximately half of potable water in urban areas is used for irrigation, and many homeowners do not know how or do not take the time to regularly adjust the programming on their controllers, these "smart controllers" have been shown to be helpful in achieving water conservation.

There are two categories in irrigation controllers: Domestic ones for gardening applications, and professional controllers for more demanding agricultural applications. While domestic (gardening) controllers can only open/close a zone (or several zones simultaneously) based on a time duration, without any feedback from the irrigation process, the professional irrigation controllers can irrigate based on volume (quantities defined in cubic metres / Gallons), get the feedback from the process, and can react to actual events happening during the process. For example, a professional controller calculates the actual flow rate running in the system when a specific zone is operated, compare this to its database, and adjust the irrigation process if deviation from the zone's flow rate is detected; This mechanism is called "Flow monitoring", and can prevent irrigation when a burst is occurring in the main line or in the zone's hydraulic components. The controller can also alert the operator locally via its interface, or remotely by sending an SMS or a message to a central control.

The use of smart professional irrigation controllers is constantly increasing as the shortage in water does.

Chemigation Valve

A chemigation valve is an apparatus designed to protect water supplies from agricultural chemicals used during chemigation, the application of chemicals such as fertilizers and pesticides through irrigation water. Most chemigation valves consist of a spring-loaded check valve, a low pressure drain, an air and vacuum relief valve, and an injection port for introducing the chemicals downstream of the check valve. Many chemigation valves also have a 4 inch inspection port so that a person can reach inside and feel if the check valve is still functional. Some governments require the use of two chemigation valves installed in series if hazardous chemicals are to be injected.

Texas State law Chapter 344.73 in Subchapter D - 1.(b) "An irrigation system which adds any chemical is considered to be a 'high health hazard'. Such an irrigation system must not be connected to any potable water supply except through a reduced pressure principle backflow prevention assembly."

Farmers and Workers

There have been many studies of farmers with the goal of determining the health effects of pesticide exposure.

The World Health Organisation and the UN Environment Programme estimate that each year, 3 million workers in agriculture in the developing world experience severe poisoning from pesticides, about 18,000 of whom die. According to one study, as many as 25 million workers in developing countries may suffer mild pesticide poisoning yearly.

Organophosphate pesticides have increased in use, because they are less damaging to the environment and they are less persistent than organochlorine pesticides. These are associated with acute health problems for workers that handle the chemicals, such as abdominal pain, dizziness, headaches, nausea, vomiting, as well as skin and eye problems.

Additionally, many studies have indicated that pesticide exposure is associated with long-term health problems such as respiratory problems, memory disorders, dermatologic conditions, cancer, depression, neurological deficits, miscarriages, and birth defects. Summaries of peer-reviewed research have examined the link between pesticide exposure and neurologic outcomes and cancer, perhaps the two most significant things resulting in organophosphate-exposed workers.

According to researchers from the National Institutes of Health (NIH), licensed pesticide applicators who used chlorinated pesticides on more than 100 days in their lifetime were at greater risk of diabetes. In a paper appearing in the May, 2008, issue of the American Journal of Epidemiology, researchers said the associations between specific pesticides and incident diabetes ranged from a 20 percent to a 200 percent increase in risk. New cases of diabetes were reported by 3.4 percent of those in the lowest pesticide use category compared with 4.6 percent of those in the highest category. Risks were greater when users of specific pesticides were compared with applicators who never applied that chemical.

The Consumers

There are concerns that pesticides used to control pests on food crops are dangerous to people who consume those foods. These concerns are one reason for the organic food movement. Many food crops, including fruits and vegetables, contain pesticide residues after being washed or peeled. Chemicals that are no longer used but which are resistant to breakdown for long periods may remain in soil and water and thus in food.

The United Nations Codex Alimentarius Commission has recommended international standards for Maximum Residue Limits (MRLs), for individual pesticides in food.

In the EU, MRLs are set by DG-SANCO. In the US, levels of residues that remain on foods are limited to tolerance levels that are established by the US Environmental Protection Agency and are considered safe. The EPA sets the tolerances based on the toxicity of the pesticide and its breakdown products, the amount and frequency of pesticide application, and how much of the pesticide (i.e., the residue) remains in or on food by the time it is marketed and prepared.

Tolerance levels are obtained using scientific risk assessments that pesticide manufacturers are required to produce by conducting toxicological studies, exposure modelling and residue studies before a particular pesticide can be registered, however, the effects are tested for single pesticides, and there is little information on possible synergistic effects of exposure to multiple pesticide traces in the air, food and water.

A study published by the United States National Research Council in 1993 determined that for infants and children, the major source of exposure to pesticides is through diet. A study in 2006 measured the levels of organophosphorus pesticide exposure in 23 school children before and after replacing their diet with organic food (food grown without synthetic pesticides). In this study it was found that levels of organophosphorus pesticide exposure dropped dramatically and immediately when the children switched to an organic diet.

To reduce the amounts of pesticide residues in food, consumers can wash, peel, and cook their food; trim the fat from meat; and eat a variety of foods to avoid repeat exposure to a pesticide typically used on a given crop. Consumers can also buy food that is grown organically, though even organic food may have traces of pesticides.

Strawberries and tomatoes are the two crops with the most intensive use of soil fumigants. They are particularly vulnerable to several type of diseases, insects, mites, and parasitic worms. In 2003, in California alone, 3.7 million pounds (1,700 metric tons) of metam sodium were used on tomatoes. In recent years other farmers have demonstrated that it is possible to produce strawberries and tomatoes without the use of harmful chemicals and in a cost effective way.

General Public

Exposure routes other than consuming food that contains residues, in particular pesticide drift, are potentially significant to the general public.

The Bhopal disaster occurred when a pesticide plant released 40 tons of methyl isocyanate (MIC) gas, a chemical intermediate in the synthesis of some carbamate pesticides. The disaster immediately killed nearly 3,000 people and ultimately caused at least 15,000 deaths.

In China, an estimated half million people are poisoned by pesticides each year, 500 of whom die.

Children have been found to be especially susceptible to the harmful effects of pesticides. A number of research studies have found higher instances of brain cancer, leukaemia and birth defects in children with early exposure to pesticides, according to the Natural Resources Defence Council. Often used for ridding school buildings of rodents, insects, pests, etc., pesticides only work temporarily and must be reapplied. The poisons found in pesticides are not selectively harmful to just pests and in everyday school environments children (and faculty) are exposed to high levels of pesticides and cleaning materials. "No testing has ever been done specifically pertaining to threats among children"

Peer-reviewed studies now suggest neurotoxic effects on developing animals from organophosphate pesticides at legally tolerable levels, including fewer nerve cells, lower birth weights, and lower cognitive scores. The United States Environmental Protection Agency finished a 10 year review of the organophosphate pesticides following the 1996 Food Quality

Protection Act, but did little to account for developmental neurotoxic effects, drawing strong criticism from within the agency and from outside researchers.

Some scientists think that exposure to pesticides in the uterus may have negative effects on a fetus that may manifest as problems such as growth and behavioural disorders or reduced resistance to pesticide toxicity later in life.

A new study conducted by the Harvard School of Public Health in Boston, has discovered a 70% increase in the risk of developing Parkinson's disease for people exposed to even low levels of pesticides.

A 2008 study from Duke University found that the Parkinson's patients were 61 percent more likely to report direct pesticide application than were healthy relatives. Both insecticides and herbicides significantly increased the risk of Parkinson's disease.

One study found that use of pesticides may be behind the finding that the rate of birth defects such as missing or very small eyes is twice as high in rural areas as in urban areas. Another study found no connection between eye abnormalities and pesticides. In the USA, increase in birth defects is associated with conceiving in the same period of the year when agrichemicals are in elevated concentrations in surface water.

Pyrethrins, insecticides commonly used in common bug killers, can cause a potentially deadly condition if breathed in.

Continuing Development

Pesticide safety education and pesticide applicator regulation are designed to protect the public from pesticide misuse, but do not eliminate all misuse. Reducing the use of pesticides and choosing less toxic pesticides may reduce risks placed on society and the environment from pesticide use. Integrated pest management, the use of multiple approaches to control pests, is becoming widespread and has been used with success in countries such as Indonesia, China, Bangladesh, the US, Australia, and Mexico. IPM attempts to recognise the more widespread impacts of an action on an ecosystem, so that natural balances are not upset.

New pesticides are being developed, including biological and botanical derivatives and alternatives that are thought to reduce health and environmental risks. In addition, applicators are being encouraged to consider alternative controls and adopt methods that reduce the use of chemical pesticides.

Pesticides can be created that are targeted to a specific pest's life cycle, which can be environmentally more friendly. For example, potato cyst nematodes emerge from their protective cysts in response to a chemical excreted by potatoes; they feed on the potatoes and damage the crop. A similar chemical can be applied to fields early, before the potatoes are planted, causing the nematodes to emerge early and starve in the absence of potatoes.

The Alternatives

Alternatives to pesticides are available and include methods of cultivation, use of Biological controls, such as pheromones and microbial pesticides, and genetic engineering, and methods of interfering with insect breeding. These methods are becoming increasingly popular and often are safer than traditional chemical pesticides. In addition, EPA is registering reduced-risk conventional pesticides in increasing numbers.

Cultivation practices include polyculture (growing multiple types of plants), crop rotation, planting crops in areas where the pests that damage them do not live, timing planting according to when pests will be least problematic, and use of trap crops that attract pests away from the real crop. In the US, farmers have had success controlling insects by spraying with hot water at a cost that is about the same as pesticide spraying.

Release of other organisms that fight the pest is another example of an alternative to pesticide use. These organisms can include natural predators or parasites of the pests. Biological pesticides based on entomopathogenic fungi, bacteria and viruses cause disease in the pest species can also be used.

Interfering with insects' reproduction can be accomplished by sterilising males of the target species and releasing them, so that they mate with females but do not produce offspring. This technique was first used on the screwworm fly in 1958 and has since been used with the medfly, the tsetse fly, and the gypsy moth. However, this can be a costly. time consuming approach that only works on some types of insects.

Another alternative to pesticides is the thermal treatment of soil through steam. Soil steaming kills pest and increases soil health.

In India, traditional pest control methods include using Panchakavya, the "mixture of five products." The method has recently experienced a resurgence in popularity due in part to use by the organic farming community.

Some evidence shows that alternatives to pesticides can be equally effective as the use of chemicals. For example, Sweden has halved its use of pesticides with hardly any reduction in crops. In Indonesia, farmers have reduced pesticide use on rice fields by 65% and experienced a 15% crop increase.

18

Chemigation

Chemigation is injecting a chemical into irrigation water and applying it (chemical) through the system to the crop or field. Applying chemicals with irrigation water is not new but has been used for years with furrow irrigation and anhydrous ammonia or with flood irrigation on rice and different chemicals. Chemigation is used mostly in conjunction with sprinkler or drip irrigation systems. When using irrigation to apply chemicals, a quarter-inch application of water per acre is equivalent to applying 6,790 gallons of water (with the chemical) on each acre.

The earliest work using chemicals through sprinklers was injecting a fertilizer into the system and applying it with irrigation water (fertigation). Herbigation followed when herbicides were applied through a sprinkler system, generally centre pivots. The system was used as a long spray boom. Reduced amounts of water are used with herbicides. Other chemicals then started to be used – insecticides for insectigation, fungicides for fungigation, and nematicides for nematigation. All the terminology was combined and the term chemigation evolved.

Chemical injection is not just putting chemicals into an operating irrigation system and applying it to a field. When chemicals are injected into a closed pressurised system (centre pivot, linear, drip, or other type sprinkler or irrigation system) there are precautions that must be taken. Most safety equipment used on a chemigation unit is to prevent chemical contamination of water sources and loss or dilution of the chemical itself.

Water-application rate should not exceed the soil's infiltration rate.

There are advantages and disadvantages of applying chemicals through an irrigation system. Consider these pros and cons carefully before using chemical injection methods.

The following discussion concerns centre pivots, but the chemigation method can be adapted to other types of systems.

Advantages of Chemigation

1. *Uniformity of Application:* Centre pivots have good uniformity built into them; check all systems before applying chemicals. However, factory design usually gives excellent distribution of water. If a system lacks good uniformity, check the nozzling.
2. *Prescription Applications:* Applying the chemicals with a measured amount of water puts the chemical at the desired location in the soil.
3. *Economics:* Applying chemicals through chemigation is often less expensive than are conventional application methods.
4. *Timeliness:* Chemicals can still be applied when other methods could not be used because of wetness or other limiting conditions.
5. *Incorporation and Activation:* Chemicals are incorporated and activated when applied with the large volumes of water, even from light irrigations.
6. *Reduced Compaction:* Soil compaction is in a controlled traffic pattern. Compaction from equipment takes place at the drive towers where, with conventional spray equipment, it is over a larger area of the field.

7. *Crop Damage:* Clearance by the pivot helps prevent crop damage that would occur from conventional ground-application equipment.
8. *Operator Hazards:* Because an operator is not riding on the system, there is less contact with the chemical (due to drift, frequent filling, or other exposure).

Disadvantages of Chemigation

1. *Higher Management:* Chemical application requires skill in calibration and knowledge of the irrigation and chemical-injection equipment.
2. *Additional Equipment:* Injection equipment, safety devices, and storage equipment are needed items.
3. *Environmental Hazards:* Contamination of water sources is possible if the proper safety equipment is not used. Run-off from chemically laden water to other areas is possible. Personal injury could occur if the field is not posted and warning signs are not posted around the well when chemicals are being used.
4. *Increased Application Times:* Under centre pivots, application time might be longer than some conventional methods. This also allows for more interference by unfavourable weather conditions.
5. *Unnecessary Water Applications:* If a person is locked-in to chemigation, there may be times that applications need to be made and it is too wet or windy to apply the desired amount of water needed for proper chemical application. Water can also cause compaction. Even with light application, crusting of the surface soil may become a problem.

When any chemical is applied through an irrigation system, there are safety devices that must be used on the irrigation system and the injection equipment. The safety equipment protects against contamination of the general water supply and chemical spills from the chemical supply tank. The following are suggested to help protect the water supply and the injection chemical.

Chemigation Safety Equipment

Interlock

The irrigation pumping plant and the chemical injection pump must be interlocked or connected so that if the irrigation pumping plant stops, the chemical injection pump also stops. This prevents pumping the chemical mixture from the supply tank into the irrigation pipeline after the irrigation pumping plant stops. For internal combustion engines, the chemical injection pump can be belted to the drive shaft or an accessory pulley of the engine. Other alternatives include operating the injection equipment off the engine electrical system (12 v), or using the power source (oil or electric) of the sprinkler-system drive. In all cases, it is essential that if the irrigation water supply stops, the chemical injection also stops.

For an electrically motor-driven irrigation pump, a separate, small electric motor is usually needed to power the chemical injection pump. Controls for the two electric motors need to be interlocked so both motors stop when the motor on the irrigation pump stops. All wiring must conform to the National Electric Code. Some agricultural chemicals are flammable and require the use of explosion-proof motors and wiring. If you are injecting a pesticide, consult the label for specific information before use.

Irrigation Pipeline/Check Valve and Vacuum-Relief Valve

Check and vacuum-relief valves (anti-siphon devices) are needed in the irrigation pipeline. They keep water and/or a mixture of water and chemical from draining or siphoning back into the irrigation well and polluting the groundwater. Both of these valves are between the irrigation pump discharge and the place of chemical injection into the irrigation pipeline.

The check valve must have positive closing action and a watertight seal and must be easy to repair and maintain. Install the valve with fittings that allow easy removal for maintenance and repair. The vacuum-relief valve allows air into the pipeline when the water flow stops. This prevents the creation of a vacuum (behind the check valve) that could lead to siphoning.

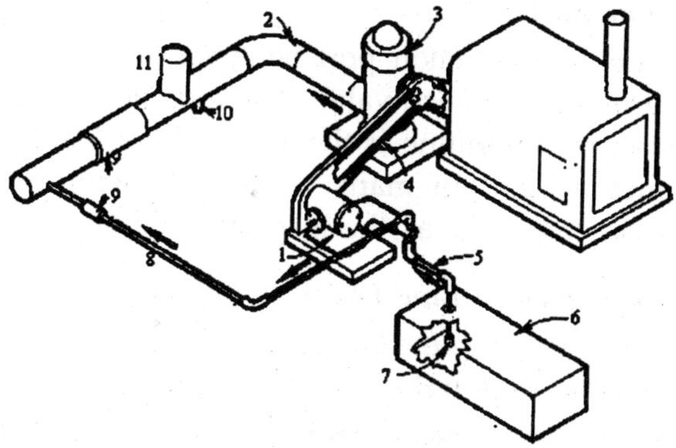

1. Chemical pump;
2. Irrigation pipeline;
3. Irrigation pump;
4. Belt driven;
5. Suction line;
6. Chemical line;
7. Suction line strainer;
8. Discharge line;
9. Check valve;
10. Automatic low pressure drain Discharge 20 ft. from well;
11. Vacuum relief valve and inspection port.

Figure: Minimum Requirements for Anti-Pollution Devices and Arrangement of Equipment for Applying Chemicals through the Irrigation System (Engine Drive)

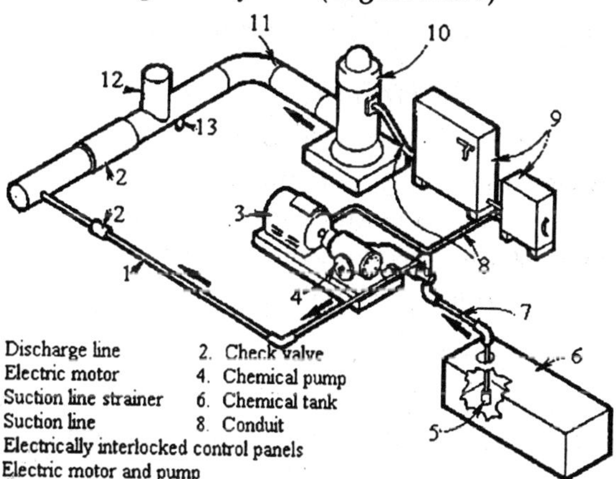

1. Discharge line
2. Check valve
3. Electric motor
4. Chemical pump
5. Suction line strainer
6. Chemical tank
7. Suction line
8. Conduit
9. Electrically interlocked control panels
10. Electric motor and pump
11. Irrigation pipeline
12. Vacuum relief valve and inspection port
13. Automatic low pressure drain Discharge 20 ft. from well

Figure: Minimum Requirements for Anti-Pollution Devices and Arrangement of Equipment for Applying Chemicals through the Irrigation System (Motor Drive).

Inspection Port

Locate an inspection port between the pump discharge and the mainline check valve. This port allows for a visual inspection to determine if the check valve leaks. Inspect at least once a year. In many cases the vacuum-relief-valve connection can serve as the inspection port.

Chemical Injection Line Check Valve

A check valve in the chemical injection line is needed for two purposes: to stop flow of water from the irrigation system into the chemical supply tank and to prevent gravity flow from the chemical supply tank into the irrigation pipeline after an unexpected shutdown. If this check valve were omitted and the injection pump stopped, irrigation water could possibly flow back through the chemical line into the chemical supply, tank, overflow the tank, and cause a spill around the irrigation well. The chemical then may eventually move down through the soil to the groundwater. It would be desirable for this check valve to have a minimum opening (cracking) pressure of 10 psi to prevent gravity flow from the chemical tank, through the injection pump, and into the irrigation pipeline. It should be constructed of chemically resistant materials.

Hoses, Clamps, and Fittings

Keep all hoses, clamps, and fittings in good repair and inspect at least annually. All components in contact with the chemical or chemical mixture (from the strainer to the point of injection on the irrigation pipeline) should be constructed of chemical-resistant materials.

Chemical Suction Line Strainer

A strainer on the chemical suction line is necessary to prevent clogging or fouling of the injection pump, check valve, or other equipment.

Low-Pressure Drain

An automatic low-pressure drain should be placed on the bottom side of the irrigation pipeline. In the event that the mainline check valve leaks slowly, the solution drains away from rather than into the well. The drain should discharge at least 20 feet from the well, and the flow should be directed away from the well. A hose or pipe may be needed to direct the discharge from the drain to the desired distance.

The preferred location of this drain is between the irrigation pump and the mainline check valve. With some systems, placement of this valve may be more feasible downstream of the mainline check valve; however, it should always be located on the irrigation pipeline before the point of chemical injection.

Additional Protection Solenoid Valve

A normally closed solenoid valve can be electrically interlocked with the power unit or motor driving the injection pump. This valve, located on the inlet side of the injection pump, provides a positive shut-off on the chemical injection line. Therefore, neither the chemical nor the water could flow in either direction if the chemical pump is stopped.

Management Practices for Chemigation

Flushing Injection System

Flush the injection system with clean water after use to prevent accumulation of precipitates and contamination of the equipment.

Flushing Irrigation System

After injection is completed, operate the irrigation pump for at least 10 minutes to flush the chemical from the irrigation system. If the irrigation system quits automatically, flush the system as quickly as possible after the shutdown is discovered, and extend the flushing period to a minimum of 30 minutes.

Monitoring

Periodic monitoring of the irrigation system and chemical injection equipment helps assure proper operation during any chemical application.

Agitation

Some form of mechanical agitation should be used where agitation is required. Paddle-wheel agitation at 60 rpm may work the best.

Calibration Check

You should check the pivot for application uniformity and overall flow rate to make sure the pivot system is performing as specified. A uniformity coefficient of 77 per cent minimum is expected.

Drive Units

It may be desirable to purchase high-speed drive units on a new system or to convert an old system so you can make lighter applications of water (where desirable with some chemicals).

Chemical Compatibility

Check compatibility of the chemical with the water supply to assure there will not be precipitate that could clog nozzles on the system.

End Guns

Check the uniformity and application under the end gun and shut it off if it is not desirable. Uniformity should match the system; however, in most cases it is recommended to shut off the end gun during injections.

Chemical Labels

Follow label directions on all chemicals when injecting them. If a chemical is not specifically labelled for use through centre pivot or for chemigation and a failure occurs, the user probably is liable.

Run-off

Manage the irrigation system to prevent run-off of the water-chemical mixture. If run-off does occur within the field, take precautions to prevent the run-off from leaving the field where a chemical is being applied. With a given sprinkler package on a centre pivot, reducing the size (or depth) of the application (i.e., making a faster revolution) reduces the potential for run-off.

Application to Surface Water

Avoid application of chemicals on those fields with permanent or semi-permanent surface water areas. Such application may affect wildlife, other non-target plants and animals, or groundwater quality.

Chemigation Pumps

There are three injection units used for chemigation. The two mechanical units are piston pumps and diaphragm pumps. The third unit is a venturi metre that creates a suction from velocity changes through the throat of the metre.

The piston and the diaphragm pumps can be powered by belt drive or electric motor. Both pumps can be adjusted for flow rate within a designed range. After adjusting the calibration on a pump, always recheck the calibration with actual measurements of material used or supplied. Also make sure all components in the pumps, pistons, seals, hoses, and couplings are resistant to the chemicals they are being used with, and always clean and flush the pump after use.

The venturi metre is a tube with a reduced diameter in the throat, creating a vacuum that pulls the chemical into the water stream. Venturi metres require a constant water supply from an external water source or a bypass with a small booster pump of some water from the system.

When using a venturi, an additional pump or booster pump must be used for steady flow (through the venturi) at a higher pressure than the pivot. A small valve is used on the suction line to regulate the injection rate. Any variations in flow rate from the water supply that creates the vacuum change the rate of injection.

All three pumps are satisfactory for injection of chemicals; calibrate each one for the volume to be injected. Clean carefully after use. If chemicals stay in the pumps, the useful life is short and there are problems from failure of seals, hoses, or other mechanical parts. No pump is any better than the time taken to calibrate it correctly. Diaphragm pumps are more popular because of ease of calibration, lack of external leaks, and maintenance of calibration.

Calibration

Accurate calibration of injection equipment is essential for safe, proper application. Small differences in injected material make large differences in the total amount of chemical applied and could cause failure or over application.

Injector Calibration

450 gallons/minute = 1 acre-inch/hour

27,00 gallons = 1 acre-inch

1. Determine total acres to be irrigated.
2. Amount of material to be applied per acre.
3. Multiply step 1 by step 2 = total material needed.
4. Desired revolution time.

Divide step 3 by step 4 = injection rate (units/time)

To check calibration, use a stopwatch and container of known-volume graduations. To determine the amount of material per minute, use the following formulas:

Gallons/hours x 63.09 = ml/minute

If using a millilitre graduated cylinder:

Gallons/hours x 2.133 = oz/minute

If using an ounce graduated cylinder:

Use 32 per cent Urea Ammonium Nitrate to apply 10 pounds N through a 1320-foot centre pivot covering 125.5 acres at 900 gallons per minute. The fertilizer goes on with 0.50 inches of water.

32 per cent Urea Ammonium Nitrate weighs 11.06 lb/gal and has 3.54 lb. N/gal

10 / 3.54 = 2.82 gallons of fertilizer

125.5 x 2.82 = 353.9 total gallons of fertilizer solution

It takes 31.3 hours to apply 0.5 inches on the whole circle.

353.9 gallons / 313 hours = 11.3 gal/hour injection rate

For calibration, 11.3 x 2.133 = 24.12 oz/min. of fertilizer solution.

Management Considerations for Fertilizer Use

Of the chemicals possible for use in chemigation, fertilizers are the best suited for commercial use in Mississippi. Methods for applications of herbicides, fungicides, and insecticides are advancing rapidly and should be developed for commercial use in the next few years.

It must be noted that fertigation is not foliar feeding. With fertigation, almost 100 per cent of the applied fertilizer ends up in the soil. This is in contrast to foliar feeding where the fertilizer with small amounts of water is applied to the plant foliage where it remains until it is used by the plant or washed off by rain or irrigation.

When using fertilizers with an irrigation system, at least four questions should be considered.

Are Additional Plant Nutrients Needed?: The decision to apply post-plant fertilizers is in some ways the most difficult aspect of chemigation. The decision is always somewhat subjective, and it can be difficult to collect data to aid in making the decision. With cotton and corn, nitrogen is the nutrient most likely needed in a post-plant application. Preplant application of phosphorus and potassium is almost always adequate for a complete growing season.

A decision to apply nitrogen after planting may be based on a history of yield responses with post-plant nitrogen applications by more conventional methods, the appearance of deficiency symptoms, or the results of tissue testing for nitrogen. Tissue testing for nitrogen has had poor acceptance in Mississippi because water stress interferes with the test results. Properly irrigated crops should be less subject to this complication.

Is Selected Material Suitable for Application with an Irrigation System?: The simplest fertilizer materials for water injection are true solutions such as urea ammonium nitrate (UAN) at 32 per cent N by weight. The chemical forms of these nutrients are compatible with typical irrigation-water chemistry and do not damage foliage in the diluted form. They have good soil mobility so that they will be moved into the rooting zone with the irrigation water.

Anhydrous ammonia (NH_3) can react with irrigation waters that are high in carbonate and bicarbonte causing precipitation of lime from the water. This can contribute to clogging of the system. Anhydrous is also more volatile than other forms of nitrogen, which will result in a greater loss of nitrogen to the atmosphere. Also, many sprinkler companies don't recommend injecting anhydrous through the system.

Some forms of high-grade soluble phosphate materials do not react with water to form large amounts of precipitates. These can be applied with water. Phosphates present another kind of problem, because they are not mobile in the soil and stay close to their point of application. With sprinkler-applied water, the phosphate stays near the soil surface where there are few roots. The result is that, even though the material can be applied to the soil by the irrigation system, they have a limited plant availability.

Is Material Economical?: There are several forms of N, P, and K that are chemically well suited for irrigation application. For example, some forms of calcium nitrate and calcium nitrate solutions are compatible with sprinkler injection and have good plant availability. However, their cost is much higher than other nitrogen materials (particularly UAN solutions) that work as well. Cost and availability of the material to be applied are important considerations to the producer.

Is the Operation Economical?: Equipment and operating costs of adding chemical injection to an existing sprinkler system are minimal compared to costs associated with conventional ground applications. Unfortunately, many Mississippi fields are only partially covered by irrigation systems, and conventional application equipment must be available for the non-irrigated parts of the fields. Maintaining conventional application equipment for non-irrigated portions of a farm may detract from the economics of water application of fertilizers or other chemicals.

Water Application

Once a grower has decided to apply a fertilizer through an irrigation system, he should consider how much water to use in the application. Sprinkler irrigation systems are designed to apply water uniformly to the soil, but small differences in soil infiltration rate within a field may result in movement of water from one part of a field to another. This redistribution of the water is not just the wholesale run-off from a large area, but the redistribution of water within relatively small areas of the field. The result is uneven amounts of available water in the soil, and these show up as hot spots or areas where rank growth occurs. The same uneven distribution of nutrients can occur. Distribution is better if the fertilizer is applied with smaller amounts of water (1/2 inch or less).

Water-application rate should not exceed the soil's infiltration rate.

Summary

Chemical application with irrigation systems is a safe and effective practice — provided you manage properly and use precautions. Regulations adopted in the future at the Federal, State, or local level may change requirements for chemical application. Keep abreast of future technology developments and regulatory requirements to assure that effective chemical applications are made and that adequate protection is provided.

Chemigation Safety Measures

Chemigation is the process of applying an agricultural chemical (fertilizer or pesticide) to the soil or plant surface with an irrigation system by injecting the chemical into the irrigation water. Depending on the type of agricultural chemical being applied, chemigation may be referred to as fertigation, herbigation, insectigation, fungigation, etc.

Chemigation, like other chemical application methods, has its advantages, limitations and risks that a farm manager must consider in deciding which method of application is the best choice. The greatest risk of chemigation is the potential for accidental backflow of all or part of the chemical into the irrigation water source if the system is not properly set up, operated and maintained.

Chemigation can be an effective and safe way of applying certain agricultural chemicals to some irrigated crops if the proper irrigation system and anti-pollution safety devices to protect the water source are utilised. Application of a portion of a crops nitrogen requirement with irrigation water is a recognised best management practice to reduce nitrate leaching losses by both University of Minnesota and Minnesota Department of Agriculture crop specialists for high nitrogen use crops grown on coarse textured soils.

The 1987 Minnesota Legislature mandated that chemigation regulations and a permit programme be developed for pesticide application and expanded the law in 1989 to include fertilizers. Both laws granted administration responsibilities to the Minnesota Department of Agriculture (MDA), Agronomy Services. The MDA initiated the pesticide chemigation regulations in January 1989 and adopted regulations in fall of 1992 for fertilizer chemigation. The Minnesota Department of Health (MDH) also has rules associated with chemical storage tanks, chemigation systems and water wells (irrigation, potable and public water systems) that are referenced in the MDA regulations.

This publication does not discuss the protection requirements for chemigation systems connected to a potable well or public water supply system.

The purpose of this publication is to describe the general chemigation safety measures and management practices. These practices minimise the potential risk of accidentally allowing any injected chemical to flow back into the irrigation well, surface water source or to discharge onto the land where not intended, possibly creating a public health problem.

Specific details on the safety devices and measures should be obtained from the MDA before installing any chemigation equipment.

Major Pollution Risks

Chemigation has three main ways of potentially polluting irrigation water sources if safety devices are not functioning correctly. These three are:

1. The chemical in the supply tank and in the irrigation pipeline could flow or be siphoned back into the water source when the irrigation system shuts down.
2. The chemigation system could continue to inject chemical into the irrigation pipe line when the irrigation system shuts down causing the chemical solution to flow back into the water source or spill onto ground.
3. The chemigation system could shut down while the irrigation system continues to operate and force water back into the chemical supply tank causing it to overflow and spill onto the ground.

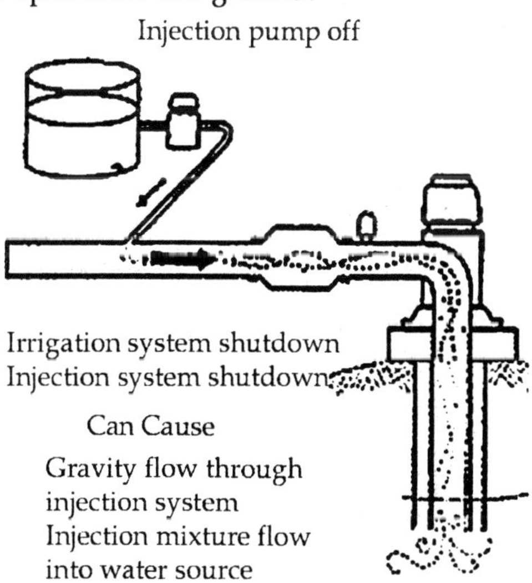

Figure: Irrigation System Shutdown / Injection System Shutdown.

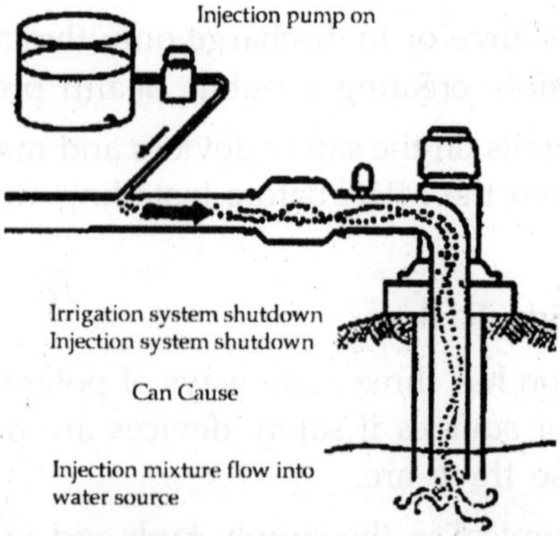

Figure: Irrigation System Shutdown / Injection System Operating

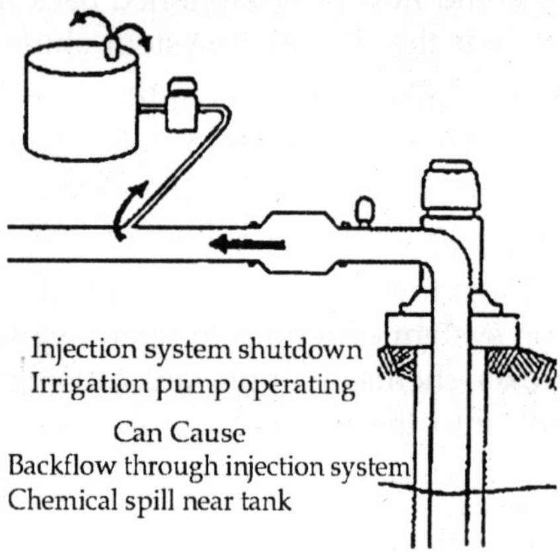

Figure: Injection System Shutdown / Irrigation Pump Operating

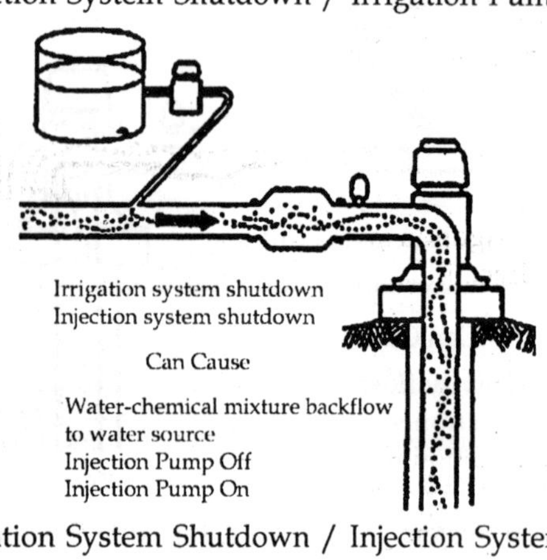

Figure: Irrigation System Shutdown / Injection System Shutdown.

The chemigation operator and farm manager/owner are responsible for seeing that these pollution risks and others are minimised through the use of proper safety devices and management measures.

Safety Measures Required by Chemigation Law

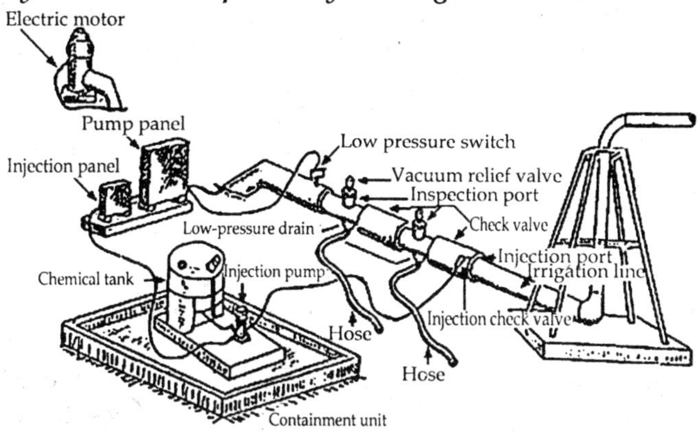

Figure: Chemigation safety equipment arrangement when applying a pesticide with an irrigation system connected to an irrigation well.

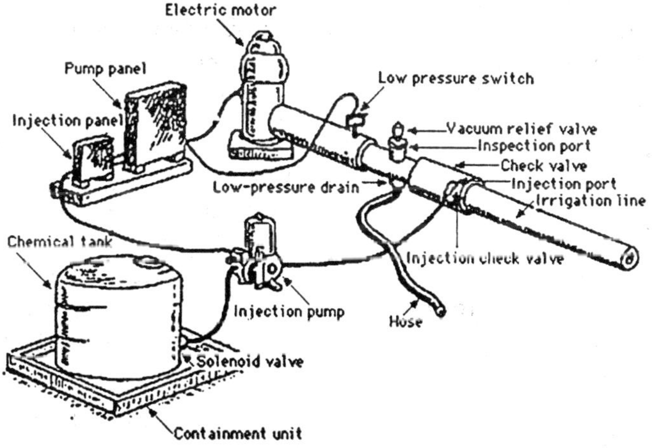

Figure: Chemigation safety equipment arrangement when applying a fertilizer with an irrigation system connected to an irrigation well.

The Minnesota Pesticide and Fertilizer Chemigation Regulations require the owner/operator of any irrigation system intending to practice chemigation to obtain a chemigation system user permit, install several safety (antipollution and safeguard) devices, comply with MDHs well separation distance rules and implement several management measures.

A chemigation user permit application form and details on safety equipment requirements are available from the Minnesota Department of Agriculture, Agronomy Services office at 90 West Plato Boulevard, St. Paul 55107 (612/297-2614). The MDA staff will inspect chemigation systems both on a routine and complaint basis.

Above figures show typical arrangements of the basic safety devices. Some requirements will vary depending on type of chemical (pesticide or fertilizer) and location of the water source. Alternative safety devices can only be used if approved in advance by MDA staff. Actual language for the MDA chemigation regulations can be found in Minnesota Rules, parts 1505.2100 - 1505.2800. MDH regulations related to well separation distances from potential contamination sources are listed in Minnesota Rules, part 4725.4450.

The basic safety devices and measures (as outlined in the MDA chemigation regulations) to protect the irrigation water source from any of the above pollution risks are as follows:

1. *Irrigation Main Pipeline Check Valve:* While injecting a pesticide, an MDA approved reduced pressure zone (RPZ) backflow preventer or two check valves in series must be provided in the main irrigation water supply pipeline of any system directly connected to an irrigation water well or a surface water source. If you intend to only inject fertilizer, a single MDA approved check valve may be used in the main pipeline.

   The check valve(s) or RPZ assembly must be located between the point of chemical injection and the irrigation water supply pump. Their main purpose is to keep the water and chemical mixture from flowing back or being siphoned back into the water source. Check valves should be installed with fittings that allow for easy removal for maintenance or repair. A check valve(s) assembly may be installed as a portable unit and moved to other permitted irrigation systems.

Each check valve assembly must contain an air vacuum relief valve and automatic low pressure release drain on the water supply side of the check valve flapper. The check valve assembly must also have an inspection port which can be easily opened to inspect the check valve flapper and the low pressure drain when the irrigation system is shut down.

The vacuum relief valve allows air to enter the pipeline when the water stops flowing. This prevents the creation of a vacuum that could cause siphoning of the water and chemical mixture down-stream of the check valve back into the water supply.

The low pressure drain must be located on the bottom of the pipeline on the supply side of the check valve and have a fully functioning drain opening of at least a 3/4 inch diameter. It must open automatically whenever the irrigation water flow stops. This provides a secondary safety backup to preventing any chemical and water mixture from entering the water source if the check valve should leak. The drain outlet must be positioned or the drainage directed to flow away from the well or surface water source during shut-down. A hose, pipe or open conduit can be used to direct the drain discharge.

Approved check valve assemblies must meet MDA design and operating standards and be certified by an independent testing laboratory. A list of currently approved check valve models can be obtained from the MDA. Check valve assemblies will be quick closing by spring action, provide a water tight seal, constructed of material resistant to corrosion or protected to resist corrosion, and easy to maintain and repair.

2. *Power Interlock:* The chemigation injection system must be interlocked with the irrigation systems power or

water supply so it will shut down anytime the irrigation system or pumping plant stops operating or the water flow is disrupted. In all cases this measure must prevent injecting chemical from the supply tank into the main irrigation pipeline after the water supply stops flowing.

If electric motors are used for both the irrigation and chemigation systems, the control panels for the two systems must be interlocked. This interlock must be set up so the injection pump motor stops whenever the irrigation system or pump stops.

Irrigation pumps driven by an internal combustion engine can be interlocked with an injection pump by being belted to the drive shaft or an accessory pulley on the engine. If the injection pump is electrically powered it should be connected to the engine's generator or electrical control system.

Some chemigation systems use flowing water or water pressure to power the injection metre or pump. In most cases these systems will stop injecting a chemical when the irrigation water supply stops flowing.

If chemical flow from the supply tank could possibly continue after shutdown, a normally closed solenoid valve should be provided in the chemical injection line, preferably on the suction side of the injection metre. The solenoid valve must be interlocked and powered by the irrigation system control panel, water supply pressure or the injector power supply.

3. *Chemical Injection Line Check Valve:* The chemical injectors discharge line/hose must contain a positive closing check valve which will not allow flow either way when the injection system is not operating. The check valve must be located between the injection metre and the point of the chemical injection into the irrigation pipeline.

   This valve should 1) stop flow of water from the irrigation system into the chemical supply tank if the injection system stops and 2) prevent gravity flow from the chemical tank into the irrigation pipeline following an unexpected shutdown. To provide two way protection the valve should have a watertight sealing check valve with a minimum opening (cracking) pressure of 10 pounds per square inch. It should also be constructed of an agricultural chemical corrosive resistant material.

If irrigation water is allowed to flow back into the chemical supply tank it could overflow the tank causing chemical to spill onto the ground. If chemical in the supply tank is allowed to flow into the irrigation pipeline by gravity or be siphoned when the irrigation system was not operating, it could create a potential danger to the crop or leak onto the ground. Then it could get into a surface or ground water source.

4. *Low Pressure Shut Down Switch:* The irrigation system must contain a low pressure shut-down switch or device with similar operating characteristics on the main pipeline that will shut down the irrigation system and the chemigation system if the operating pressure drops to an unsatisfactory level for proper agricultural chemical distribution.

5. *Chemical Supply Tank:* The chemigation supply tank must not be located closer to an irrigation well than the distance specified in the MDH rules, chapter 4725. And, it must be safeguarded according to the MDA specifications described in the following discussion. The separation distance from a surface water source must likewise be no less than that specified for an irrigation well unless other state/federal regulations are more applicable.

    - The chemigation supply tank must be housed in a secondary containment (dike) unit if the tank storage meets at least two of the following conditions: a) the supply tank has a rated capacity of more than 1500 gallons, b) the tank is located within 100 feet of a water supply, and c) the supply tank storage is located at the site for more than 30 consecutive days.

- The minimum required capacity for a secondary containment unit is 125 per cent of the tank capacity (110% if under a roof). Its walls and base may be made of ferrous metal, reinforced concrete, solid reinforced masonry, synthetic lined earth or prefabricated metal or synthetic materials. Synthetic liners must have a minimum thickness of 30 mils.
- The unit must be leakproof and built to withstand the hydrostatic pressure from the release of a full tank. The walls or base must not contain a drain. Design specifications for some types of units are described in Mid-West Plan Service Bulletin 37, *"Designing Facilities for Pesticide and Fertilizer Containment."* This bulletin is available at county extension offices.

  The chemical supply tank must be constructed from materials such as fibreglass, polyethylene or stainless steel that are resistant to the chemical being stored and resistant to degradation by sunlight. If not contained in a secondary unit, the tank should be located and landscaped so if a leak develops it will direct any leakage away from entering the water source. The tank should also be protected from damage from farm machinery and livestock.

6. *Posting of Field:* All sites being treated with a pesticide through the irrigation system must be posted with signs during the complete chemigation treatment.

   Signs must contain the signal word from the pesticide label, name of the pesticide, date of treatment and re-entry date as described by the pesticide label. Example of a sign can be obtained from the MDA.

   Signs must be posted at usual points of entry and at property corners immediately adjacent to public transportation routes or other public or private property. Signs must be placed no greater than 100 feet apart for a field that is located adjacent to a public area such as a park, school or residential area. If more restrictive instructions for posting are described on the label, those restrictions must be completely followed.

Additional Protection Measures

In addition to the safety equipment and measures previously described, several other devices and management measures can improve chemigation operation and safety to the environment. These include:

1. *Hoses, Clamps and Fittings:* All components in contact with the chemical mixtures should be constructed of materials that are chemically resistant and resistant to sunlight degradation. The pressure rating of all components should be adequate to withstand all operating pressures. Hoses and fittings should be protected from mechanical damage.
2. *Injection Line Strainer:* Install a strainer on the chemical suction line/hose to remove foreign materials that could plug or damage the injection metre/pump or chemical injection line check valve.
3. *Injection Port Location:* When possible, locate the port for chemical injection higher than the chemical supply tank but lower than the lowest sprinkler outlet to prevent siphoning from the tank. In all cases the injection port must be located downstream from the main pipeline check valve.
4. *Injection Line Flow Sensor:* An injection line flow sensor installed just upstream from the chemical injection line check valve and interlocked with the injection device can be used to shut down the injection system if flow in the injection line ceases. This safety measure would prevent continuous operation after a line/hose ruptures or disconnects, injection device loses prime or fails, supply tank is emptied or injection port becomes plugged. The flow sensor could also be interlocked

with the irrigation system to shut down the whole system if injection line flow stops.

5. *Two-way Interlock:* A two-way interlock between the irrigation system and the injection system can stop either system if the other system also stops. This interlock will eliminate untreated areas in the field by stopping the irrigation pump and sprinkler system if the injection system stops or malfunctions. The interlock can be done electrically or with the use of a flow sensor on the discharge side of the chemical injection device. When there is no flow in the injection line, the irrigation system and pumping plants will shut down.
6. *Solenoid Valve:* For added safety, a normally closed solenoid valve could be installed on the suction side of the injection device to provide a positive shut down on the injection line when not in use. The solenoid valve must be interlocked with the injection device power supply to open or close properly.
7. *Bleed Valve:* A bleed valve, located upstream and next to the injection line check valve can be used to relieve locked-in pressure in the chemical injection line any time the line is to be disconnected. This will prevent the operator from being sprayed with the chemical in the line during line removal.
8. *Calibration Equipment:* A calibration tube or in-line flow metre installed on a chemigation system can provide an easy way to measure the rate of flow of the chemical being injected into the irrigation system. A calibration tube is typically a clear tube with markings in millilitres or fluid ounces and used with a stop watch to measure the flow rate. The tube should be placed on the suction side of the injection device with the necessary valves and fittings so the injection rate can be checked during a chemigation. An in-line flow metre can be installed in either the suction or discharge side of the injection

device. Its markings are typically expressed in flow units of volume per time.

9. *Injection Metre/Pump:* Locate the injection metre or pump within the chemical supply tank containment unit when available and possible.
10. *Portable Chemigation System and Chemical Supply Tank:* Install the chemigation injection metre/pump and chemical supply tank onto a portable trailer or truck. Construct a secondary containment unit of appropriate size on the bed of the trailer or truck.
11. *Chemigation System Location:* When developing a new irrigation system locate the irrigation well at least 150 feet from the chemigation system, chemical supply tank, injection port, power interlock controls, etc.

Management Practices

1. *Review Operation of Irrigation System:* Periodically observe the irrigation systems water distribution pattern and conduct a water distribution test of the spray pattern Remember that the uniformity of the chemical distribution will be no better than the distribution of the water. Adjust the irrigation system (such as the end gun) to prevent spray going beyond the boundaries of the target field. Shut the irrigation system down if wind will carry chemical drift off target. The irrigation system should be managed to prevent run-off or deep percolation of the water-chemical mixture.

   Do not chemigate in areas containing wetlands and other surface water bodies. Do not apply any pesticide that is not labelled for use in an irrigation system. Such applications are illegal and may adversely affect wildlife, non-target plants, and water quality.

2. *Inspect Safety and Anti-pollution Equipment Before Each Use:* Inspect all components of the chemigation and irrigation system before each use. Components not working at the time of inspection should be repaired or replaced before chemigating. Routine inspections

should minimise the potential for failure of any component during chemigation. A record of inspection dates must be kept for MDA.

Follow the procedures listed below to inspect the irrigation pipeline check valve, low pressure drain, injection line check valve, low pressure switch and the power interlock. RPZ backflow preventers and some other types of check valves will require a different approach to inspection. Contact MDA staff for directions on inspecting if assistance is needed.

- Connect the chemigation system to the irrigation system but leave the chemical injection line/hose disconnected from the injection port check valve.
- Start the irrigation pump and pressurise the irrigation system to its normal operating pressure.
- Observe the injection line check valve to see if any water is leaking back out the inlet side of the check valve. There should be no leakage observed when the irrigation system is operating or shut down.
- Connect the chemical injection hose to the injection check valve and start up the chemigation system. The chemigation system should be operated only with clean water.
- Turn down the main pipeline control valve (reducing the operating pressure) until the low pressure switch shuts down the irrigation system. The pressure switch should be set to cause the irrigation pump and system to shut down when the normal operating pressure has been reduced by 25 to 50 per cent. If no flow control valve is present, shut power off to the pump and/or irrigation system.
- Immediately after shut down check to see if any water is flowing from the low pressure drain(s). Some drainage for a short period of time after shut down is normal; then drainage should stop.

- Check to see if the chemigation injection device has stopped operating. This device should stop when the irrigation system and pump shuts down. If the chemigation system has an agitation system this unit does not have to shut down when the injection device stops.
- Open the inspection port at the main pipeline check valve assembly after the low pressure drain has stopped flowing. Inspect for any leakage from the check valve flapper. No leakage should be observed coming from the downstream flapper. Also check for proper functioning of the flapper valve assembly.

3. *Filling of Supply Tank/Mixing of Agricultural Chemicals:* Direct suppliers to closely monitor the delivery of products to a supply tank. Make sure that they first check the condition of the supply tank and plumbing before filling. Fill the supply tank to no more than 95 per cent of capacity. Monitor the inventory of product contained in the supply tank between chemigations in order to determine if the supply tank is leaking.

   Triple-rinse pesticide containers at time of use and add the rinse water into the supply tank. Rinse over the opening of the supply tank to minimise risk of spilling on the ground.

4. *Keep Chemigation Site Uncontaminated:* To facilitate safe monitoring of the chemigation operation, do not allow the irrigation system to spray water and chemical into the chemigation equipment area. This may mean plugging a couple nozzles on the irrigation system near the chemigation site.

5. *Calibration:* Accurate calibration of chemical injection device is essential for proper application. Periodically recheck the calibration setting of the injection device. Follow calibration procedures described by the chemical label, chemigation equipment manufacture or Minnesota Extension Service. Minor differences in injection rate over an extended period can cause too

high or low a chemical application. This may produce unsatisfactory results or cause potential pollution or crop damage when high applications are made.

6. *Empty Chemigation Supply Tank:* Left-over pesticide mixtures should be immediately applied to another crop or site listed on the label or removed from the supply tank and stored in an appropriate place and in a marked container for later use. The empty tank should be rinsed out and rinsing water applied to the irrigated crop or another labelled site. More specific information on storage or disposal can be obtained form MDA.

   For tanks containing left-over N solution that you're not planning to use in the next week or two, remove the N from the supply tank, or relocate the tank away from the water source, unless safeguarded according to MDA and MDH requirements.

7. *Flush Injection Equipment:* Flush the chemigation injection device, hoses and check valve with clean water after each use. Flush cleaning water into the irrigation system while it is operating so the cleaning water will be applied to the field. Clean strainer after each chemigation.

8. *Flush Irrigation System:* After a chemigation event is completed and the chemigation system is cleaned, operate the irrigation pump as long as necessary to flush the irrigation system of chemical. This may take 10 to 15 minutes for most systems.

9. *Report Accidental Spills:* If an accident occurs (regardless of size), avoid personal contamination, take action to keep potential spill to a minimum, and report the incident to the MDA immediately (1-800-422-0798) for assistance.

Bibliography

Acock, B. and Senn, T.L.: *The Effect of Humic Acids on the Growth of Tomato Plants and the Physical Properties of Sand and Sandy Loam Soils,* South Carolina Agricultural Experiment Station, Clemson University, 1964.

Aitken J.B., Acock B. and Senn, T.L.: *The Characteristics and Effects of Humic Acids Derived From Leonardite, Technical Bulletin 1015,* South Carolina Agricultural Experiment Station, Clemson University, 1964.

Anderson, D.B.: *Demand Analysis for Nitrogen Fertilizers in Saskatchewan,* Department of Agricultural Economics, University of Saskatchewan, Saskatoon, 1984.

Arditti, Joseph: *Fundamentals of Orchid Biology,* John Wiley & Sons, New York, 1992.

————: *Orchid Biology: Reviews and Perspectives,* John Wiley & Sons, New York, 1994.

Bailey, D. and Brorsen, B.W.: *Dynamics of Regional Fed Cattle Prices,* Western Journal of Agricultural Economics, 1985.

Bailey, D.V. and Brorsen, B.W.: *Price Asymmetry in Spatial Fed Cattle Markets,* Western Journal of Agricultural Economics, 1989.

Basnyat, B.B.: *Meeting the Fertilizer Challege: A Study on Nepal's Fertilizer Trade Liberalisation Programme,* Policy Analysis in Agriculture and Related Resource Management, Kathmandu, 1999.

Bayri, T., Rosaasen, K.A. and Furtan, W.H.: *Limit Pricing in the Nitrogen Fertilizer Market: An Application to the Saskatchewan Market,* Department of Agricultural Economics, University of Saskatchewan, 1986.

Bennett, J.B.: *An Econometric Simulation of the Potash Industry,* Tufts University, Thesis, 1984.

Bessler, D. and Brandt, J.A.: *Causality Tests in Livestock Markets,* Amer.J. Agr. Econ., 1982.

Bessler, D.: *An Analysis of Dynamic Economic Relationships: An Application to the U.S. Hog Market,* Canadian Journal of Agricultural Economics, 1984.

Borchardt, Glenn, Hirschfeld, S.E. Lienkaemper, J.J. McClellan, Patrick, Williams, P.L. and Wong, I.G.: *Proceedings of the Second Conference on Earthquake Hazards in the Eastern San Francisco Bay Area,* California Department of Conservation, Division of Mines and Geology Special Publication, 1992.

Borchardt, Glenn, Rice, Salem and Taylor, Gary: *Paleosols Overlying the Foothills Fault System Near Auburn,* California Division of Mines and Geology, California, 1980.

Borchardt, Glenn: *Fault Features in Soils of the Mehrten Formation,* California Division of Mines and Geology, California, 1980.

————: *The Scientific Worldview: Berkeley,* Progressive Science Institute, 1984.

Boyd, S.M. and Brorsen, B.W.: *Price Asymmetry in U.S. Pork Marketing Channel,* North Central Journal of Agricultural Economics, 1988.

Boyle, G.E.: *Modeling Fertilizer Demand in the Republic of Ireland: A Cost Function Approach,* Journal of Agricultural Economics, 1982.

Burrell, A.: *The Demand for Fertilizer in the United Kingdom,* Journal of Agricultural Economics, 1982.

California Fertilizer Association: *Soil Improvement Committee: Western Fertilizer Handbook,* Interstate Publishers, Danville, 1995.

Carman, H.F.: *The Demand for Nitrogen, Phosphorus and Potash Fertilizer Nutrients in the Western United States,* Western Journal of Agricultural Economics, 1979.

Chambers, M.J.: *"Consumers" Demand in the Long Run: Some Evidence from U.K. Data,* Applied Economics, 1993.

Crawford, J.H., Senn, T.L. and Stembridge, G.E.: *The Influence of Humic Acid Fractions on Sprout Production and Yield of the Caragold Sweet Potato,* South Carolina Agricultural Experiment Station, Clemson University, 1968.

Dante, E.T.: *Fertilizer Market Development in the Philippines, Agro-chemicals News in Brief,* FADINAP, United Nations/ ESCAP, Hegenbarth, P. and Rodricks, 1999.

Davidson, Dr. O. Wesley: *Orchid Ailments not Caused by Insects or Diseases,* Aos Bulletin, 1990.

Davies, P.: *Market Power and Cost Structure in the Saskatchewan Potash Industry,* University of Saskatchewan, Thesis, 1992.

Davis, J.F., Bennett, J.H., Borchardt, Glenn, Kahle, J.E., Rice, S.J. and Silva, M.A.: *Earthquake Planning Scenario for a Magnitude 8.3 Earthquake on the San Andreas Fault in Southern California,* California Division of Mines and Geology, California, 1982.

Denbaly, M. and Vroomen, H.: *Dynamic Fertilizer Nutrient Demands for Corn: A Cointegrated and Error-Correcting System,* American Journal of Agricultural Economics, 1993.

Dickey, D.A. and Fuller, W.F.: *The Likelihood Ratio Statistics for Autoregressive Time Series with a Unit Root,* Econometrica, 1981.

Engle, R.F. and Granger, C.W.J.: *Co-integration and Error-Correction: Representation,* Estimation and Testing, Econometrica, 1987.

Faminow, M.D. and Benson, B.L.: *Integration of Spatial Markets,* American Journal of Agricultural Economics, 1990.

Featherstone, A.M. and Baker, T.G.: *An Examination of Farm Sector Real Asset Dynamics:1910-1985,* Amer.J. Agr. Econ, 1987.

Flaig, B.: *Soil Components,* J.E.Gieseking, Ed., New York, Springer, 1975.

Frankenberger, W.T. and Sims, R.C.: *Bioremediation of Contaminated Soils,* American Society of Agronomy, 1999.

Gardner, B.L.: *The Farm-Retail Price Spread in a Competitive Food Industry,* Amer. J. Agr. Econ., 1975.

Goodwin, B.K. and Schroeder, T.C.: *Price Dynamics in International Wheat Markets,* Canadian Journal of Agricultural Economics, 1991.

Granger, C.W.J.: *Co-integrated Variables and Error-Correcting Models,* University of California, San Diego, 1983.

Green Markets: *Fertilizer Market Intelligence Weekly: Prices, Politics, Economics, Transportation,* A Pike & Fischer Publication, 1995.

Green, R., Hassan, Z. and Johnson, S.R.: *Selection of Nonnested Demand Models,* Canadian Journal of Agricultural Economics, 1995.

Griliches, Z.: *Distributed Lags, Desegregation and Regional Demand Functions for Fertilizer,* Journal of Farm Economics, 1959.

———————: *The Demand for Fertilizer: An Economic Interpretation of a Technical Change,* Journal of Farm Economics, 1958.

Gunjal, K.R., Roberts, R.K. and Heady, E.O.: *Fertilizer Demand Function for Five Crops in the United States,* Southern Journal of Agricultural Economics, 1980.

Haddock, D.D.: *Basing-Point Pricing: Competitive vs. Collusive Theories,* The American Economic Review, 1982.

Hakkio, C.S. and Morris, C.S.: *Vector Autoregressions: A User's Guide,* Federal Reserve Bank of Kansas City, Kansas City, 1984.

Heady, E.O. and Martin H. Yeh.: *National and Regional Demand Functions for Fertilizer,* Journal of Farm Economics, 1959.

Higgins, J.: *Input Demand and Output Supply on Irish Farms: A Micro-Economic Approach,* European Review of Agricultural Economics, 1986.

Houck, J.P.: *An Approach to Specifying and Estimating Nonreversible Functions,* Amer. J. Agr. Econ, 1977.

Huang, X.: *Future Directions of the Fertilizer Industry in China,* IFA Regional Conference for Asia and the Pacific, Paris, 1999.

Huq, S.: *Bangladesh: Review of its Fertilizer Development and Trade Potentials: Paper Presented at the 1999 IFA Regional*

Conference for Asia and the Pacific, Kuala Lumpur, Malaysia, 1999.

International Fertilizer Development: *Corporation, Recent Fertilizer Project Announcements Worldwide,* Muscle Shoals, Alabama, 2000.

Isherwood, K.F.: *IFA's Fertilizer Demand Forecasts, A Working Paper at the 67th IFA Annual Conference,* Manila, the Philippines, Paris, 1999.

Jeebe, M.: *Fertilizer Policy in Pakistan and Nepal,* Agro-chemicals News in Brief Special Issue, 1999.

Jixian, M.: *A Being Globalized and Marketoriented China Fertilizer Circulation System,* Manila, the Philippines, Paris, 1999.

Khristeva, L.A. and Manoilova, A.: *The Nature of the Direct Effect of Humic Acids on the Growth and Development of Plants,* Translated abstract in Commonwealth Bureau of Soils Bibliography on Humate Fertilizers, 1950.

Kimbro, W. and Waer, G.D.: *Formulations of Katy-J Multi-Plex Products,* JKT Corporation, 1995.

Kinnucan, H.W. and Forker, O.D.: *Asymmetry in the Farm-Retail Price Transmission for Major Dairy Products,* Amer. J. Agr. Econ., 1987.

Koepke, W.E.: *Structure, Behavior and Performance,* World Potash Industry, 1973.

Kononova, M.M.: *Soil Organic Matter: Its Role in Soil Formation and in Soil Fertility,* Translated From the Russian by T.Z. Nowakowski and G.A. Greenwood, 1961.

Korol, M. H.: *Pricing Behavior in the Potash Industry: The Dominant Firm Model, Limit Pricing and Price Discrimination,* University of Saskatchewan, Saskatoon, 1987.

Larue, B. and Tanguay, L.: *Regional Price Dynamics and CounterVailing Duties: Did the Canada - U.S. Hog/Pork Dispute Have a Permanent Impact?,* Not Published Article, 1997.

Lingard, J.: *Aggregate Demand for Nitrogenous Fertilizer in the United Kingdom: Another Look,* Journal of Agricultural Economics, 1971.

Louis, P.L.: *Fertilizer Supply and Trade Year 2000 Situation – Outlook for 2001,* IFA Enlarged Council Meeting, Brazil, 2000.

————: *Fertilizers and Raw Materials Supply and Supply/ Demand Balances,* Manila, the Philippines, Paris, 1999.

MacGregor, R.J.: *Impact of Policy on Bi-Lateral Trade in Wheat Between Western Canada and the U.S.,* Economic Analysis Division, Agriculture and Agri-Food Canada, 1995.

Mackinnon, J.G.: *Approximate Asymtotic Distribution Functions for Unit Root and Cointegration Tests,* Queen's University, Ontario, 1992.

————: *Critical Values for Cointegration Tests, Long-run Economic Relationships: Reading in Cointegration,* Oxford University Press, Oxford, 1991.

Marsh, J.M. and Brester, G.W.: *Intertemporal Price Adjustments in the Beef Market: A Reduced Form Analysis of Weekly Data,* Western Journal of Agricultural Economics, 1989.

Martin, J.A. and Senn, T.L.: *The Influence of Various Rates of Nitrogen and Humic Acid Derivatives on the Growth and Yield of Greenhouse Tomatoes, Research Series Number 95,* South Carolina Agricultural Experiment Station, Clemson University, 1967.

Martin, J.A., Senn, T.L., Crawford, J.H. and Moore, M.D.: *Influence of Humic and Fulvic Acids on the Growth, Yield and Quality of Certain Horticultural Crops, Research Series Number 30,* South Carolina Agricultural Experiment Station, Clemson University, 1962.

Miller, R.W. and Gardiner, D.T.: *Soils in Our Environment,* Upper Saddle River, Prentice Hall, 1998.

Mohanty, S., Peterson E.W.F. and N. Cottrell Kruse: *Price Asymmetry in the International Wheat Market,* Canadian Journal of Agricultural Economics, 1995.

Payne, G.D.: *Spatial Fertilizer Pricing in Western Canada,* Department of Agricultural Economics, University of Saskatchewan, Saskatoon, 1997.

Perman, R.: *Cointegration: An Introduction to the Literature,* Journal of Economic Studies, 1991.

Pidgeon E. and Kiwanuka, L.: *Regional Fertilizer Demand Analysis,* Agriculture and Agri-food Canada, Policy Branch, 1987.

Pierzynski, G.M.; Sims, J.T.; and Vance, G.F.: *Soils and Environmental Quality,* Boca Raton, CRC Press, 2000.

Punyawadee, V., Boyd M.S. and Faminov, M.D.: *Testing for Asymmetry Pricing in the Alberta Pork Market,* Canadian Journal of Agricultural Economics, 1991.

Quiggin, J. and Bui-Lan, A.: *The Use of Cross-Sectional Estimates of Profit Functions for Tests of Relative Efficiency: A Critical Review,* Australian Journal of Agricultural Economics, 1984.

Roberts, R.K. and Heady, E.O.: *Fertilizer Demand Functions for Specific Nutrients Applied to Three Major U.S. Crops,* Western Journal of Agricultural Economics, 1982.

Rong, M.: *The Development of Fertilizer Industry in China, Paper Presented at the International Workshop on Economic Policy Reforms and Agricultural Input Markets: Experiences,* Lessons and Challenges, South Africa, 2000.

Saxena, S.K.: *Recent Agricultural and Fertilizer Developments in India, Paper Presented at the 1999 IFA Regional Conference for Asia and the Pacific,* Kuala Lumpur, Malaysia, 1999.

Scherer, F.M.: *Industrial Market Structure and Economic Performance,* Houghton-Mifflin, Boston, 1980.

Schnitzer, M. and Poapst, P.A.: *Effects of a Soil Humic Compound on Root Initiation,* Nature, 1965.

Schroeder, T.C. and Goodwin, B.K.: *Regional Fed Cattle Price Dynamics,* Western Journal of Agricultural Economics, 1990.

Scully, Robert M.: *Should Orchids be Fertilized?,* Aos Bulletin, 1951.

Senn, T.L., Stembridge, G.E., Sims, E.T. and Crawford, J.H.: *Effect of Humic Acid Derivatives on the Conversion of Ammonium Nitrogen in Sandy Loam Soil, Research Series Number 96,* South Carolina Agricultural Experiment Station, Clemson University, 1967.

Sims, C.A.: *Money, Income and Causality*, Americam Economic Review, 1972.

Sims, E.T. Jr., Crawford, J.H., Martin, J.A. and Senn, T.L.: *A Comparison of the Quality of Several Vegetables Fertilized with Humic Acid Fractions and Inorganic Materials, Research Series Number 47*, South Carolina Agricultural Experiment Station, Clemson University, 1963.

Soedjais, Z.: *Expanding Fertilizer Production in Indonesia, Fertilizer Focus*, FMB Publications Limited, United Kingdom, 1999.

————: *The Effect of Fertilizer Subsidy Withdrawal on Agricultural Development in Indonesia*, Kuala Lumpur, Paris, 1999.

Soh, K.G. and Isherwood, K.F.: *A Review of the Current Agricultural and Fertilizer Situation*, Manila, the Philippines, Paris, 1999.

Statistics Canada: *Average Price of Selected Farm Inputs*, Catalogue, 1974.

Steinbrugge, K.V., Bennett, J.H., Lagorio, H.J., Davis, J.F., Borchardt, Glenn, R., Toppozada T., Degenkolb, H.J., Laverty, G.L. and McCarty, J.E.: *Earthquake Planning Scenario for a Magnitude 7.5 Earthquake on the Hayward Fault in the San Francisco Bay Area*, California Department of Conservation, Division of Mines and Geology Special Publication, 1987.

Sun, F. and Zhijian, Li: *Recent Developments and Prospects for Fertilizer in China*, Manila, the Philippines, Paris, 1999.

Sundaravadivelu, P.R.: *Review of Fertiliser Situation in South Asia 1998-99*, Rome, Paris, 1999.

Supply, Marketing: *Distribution and Use of Fertilizer in Indonesia Bangkok*, ESCAP, 1999.

Tamrakar, A.M.: *A Review on Promotion of Private Sector Fertilizer Trade in Nepal*, The Market - Fertilizer News and Analysis, London, 2001.

Toppozada, T.R., Bennett, J.H., Borchardt, Glenn, Richard, Saul, F., Davis J., Johnson, C.B., Lagorio, H.J. and

Steinbrugge, K.V.: *Planning Scenario for a Major Earthquake on the Newport-Inglewood Fault Zone,* California Division of Mines and Geology Special Publication 99, 1988.

Toppozada, T.R., Borchardt, Glenn, Hallstrom, C.L., Johnson, C.B., Ron, Per and Lagorio, H.J.: *Planning Scenario for a Major Earthquake on the San Jacinto Fault in the San Bernardino Area,* California Division of Mines and Geology Special Publication 102, 1993.

Toppozada, T.R., Borchardt, Glenn, L., Hallstrom C., Youngs, L.G., Gallagher, R.P. and Lagorio, H.J.: *Planning Scenario for a Major Earthquake on the Rodgers Creek Fault in the Northern San Francisco Bay Area,* California Department of Conservation, Division of Mines and Geology Special Publication 112, 1994.

Toppozada, Tousson, Borchardt, Glenn, Haydon, Wayne, Petersen, Mark, Olson, Robert, Lagorio, Henry and Anvik, Theodore: *Planning Scenario in Humboldt and Del Norte Counties, California for a Great Earthquake on the Cascadia Subduction Zone,* California Department of Conservation, Division of Mines and Geology Special Publication 115, 1995.

Tweeten, L. and Quance, L.: *Techniques for Segmenting Independent Variables in Regression Analysis: Reply,* Amer.J. Agr. Econ, 1971.

Vantassell, L. and Bessler, D.: *Dynamic Price Adjustments Between Commercial Purebred Cattle Markets,* Southern Journal of Agricultural Economics, 1988.

Ward, R.W.: *Asymmetry in Retail, Wholesale and Shipping Point Process for Fresh Fruits and Vegetables,* Amer. J. Agr. Econ, 1982.

Wolffram, R.: *Positivistic Measures of Aggregate Supply Elaricities: Some New Approaches - Some Critical Notes,* Amer.J. Agr. Econ, 1971.

Wu, Sihai: *A Review of the Fertilizer Market in Northeast Asia,* IFA Enlarged Council Meeting, Paris, 1999.

Yamawaki, H.: *Dominant Firm Pricing and Fringe Expansion: The Case of the U.S. Iron and Steel Industry, 1907-1930,* Review of Economics and Statistics, 1985.

Zapata, H.O. and Garcia, P.: *Price Forecasting with Time-Series Methods and Non-Stationary Data: Amn Application to Monthly U.S. Cattle Prices,* Western Journal of Agricultural Economics, 1990.

Zhengping, W.: *Chinese Fertilizer Industry: Its Probable Strategies after Joining WTO,* Kuala Lumpur, Malaysia, 1999.

References

Agricultural Input Markets: *Experiences, Lessons and Challenges,* Cape Town, South Africa, 2000.

National Fertilizer Development Centre: *Fertilizer Review 1998/99,* Islamabad, 1999.

National Fertilizer Secretariat: *The Review of Fertilizer Year 1998,* Colombo, 1999.

NG, S.: *Testing Homogeneity in Demand System When the Regressors are Nonstationary,* Journal of Applied Econometrics, 1995.

Northen, Rebecca: *Home Orchid Growing,* Prentice Hall, New York, 1990.

United Nations Industrial Development Organization: *Fertilizer Manual,* Kluwer Academic, Norwell, 1998.

Index

R

S

T